CW01238940

Conservation
Principles, Dilemmas and Uncomfortable Truths

Conservation
Principles, Dilemmas and Uncomfortable Truths

Edited by
Alison Richmond and Alison Bracker

V&A

Published in association with the Victoria and Albert Museum London

AMSTERDAM • BOSTON • HEIDELBERG • LONDON • NEW YORK • OXFORD
PARIS • SAN DIEGO • SAN FRANCISCO • SINGAPORE • SYDNEY • TOKYO

Butterworth-Heinemann is an imprint of Elsevier

Butterworth-Heinemann is an imprint of Elsevier
Linacre House, Jordan Hill, Oxford OX2 8DP, UK
30 Corporate Drive, Suite 400, Burlington, MA 01803, USA

First edition 2009
© 2009 The Board of Trustees of the Victoria and Albert Museum
and Alison Bracker. Published by Elsevier Ltd, in Association with
the Victoria and Albert Museum London. All Rights Reserved.

No part of this publication may be reproduced, stored in a retrieval system or transmitted in
any form or by any means electronic, mechanical, photocopying, recording or otherwise
without the prior written permission of the publisher

Permissions may be sought directly from Elsevier's Science & Technology Rights Department
in Oxford, UK: phone (+44) (0) 1865 843830; fax (+44) (0) 1865 853333; email:
permissions@elsevier.com. Alternatively you can submit your request online by visiting
the Elsevier web site at http://elsevier.com/locate/permissions, and selecting
Obtaining permission to use Elsevier material

Notice
No responsibility is assumed by the publisher for any injury and/or damage to persons or property
as a matter of products liability, negligence or otherwise, or from any se or operation of any
methods, products, instructions or ideas contained in the material herein.

British Library Cataloguing in Publication Data
A catalogue record for this book is available from the British Library

Library of Congress Cataloguing in Publication Data
A catalogue record for this book is available from the Library of Congress

ISBN: 978-0-7506-8201-5

For information on all Butterworth-Heinemann publications
visit our web site at http://books.elsevier.com

09 10 11 12 13   10 9 8 7 6 5 4 3 2 1

Working together to grow
libraries in developing countries

www.elsevier.com | www.bookaid.org | www.sabre.org

ELSEVIER   BOOK AID International   Sabre Foundation

Contents

vii   Contributors' biographies

xiii   Acknowledgements

xiv   Introduction
Alison Bracker and Alison Richmond

1   Chapter 1
Auto-Icons
Jonathan Rée

6   Chapter 2
The Basis of Conservation Ethics
Jonathan Ashley-Smith

25   Chapter 3
The Aims of Conservation
Chris Caple

32   Chapter 4
The Reconstruction of Ruins: Principles and Practice
Nicholas Stanley-Price

47   Chapter 5
Minimal Intervention Revisited
Salvador Muñoz Viñas

60   Chapter 6
Practical Ethics v2.0
Jonathan Kemp

73   Chapter 7
Conservation Principles in the International Context
Jukka Jokilehto

84   Chapter 8
The Concept of Authenticity Expressed in the Treatment of Wall Paintings in Denmark
Isabelle Brajer

100   Chapter 9
The Development of Principles in Paintings Conservation: Case Studies from the Restoration of Raphael's Art
Cathleen Hoeniger

Contents

113 Chapter 10
A Critical Reflection on Czechoslovak Conservation-Restoration: Its Theory and Methodological Approach
Zuzana Bauerová

125 Chapter 11
The Problem of Patina: Thoughts on Changing Attitudes to Old and New Things
Helen Clifford

129 Chapter 12
Archaeological Conservation: Scientific Practice or Social Process?
Elizabeth Pye

139 Chapter 13
Conservation and Cultural Significance
Miriam Clavir

150 Chapter 14
The Cultural Dynamics of Conservation Principles in Reported Practice
Dinah Eastop

163 Chapter 15
Why Do We Conserve? Developing Understanding of Conservation as a Cultural Construct
Simon Cane

177 Chapter 16
Heritage, Values, and Sustainability
Erica Avrami

184 Chapter 17
Ethics and Practice: Australian and New Zealand Conservation Contexts
Catherine Smith and Marcelle Scott

197 Chapter 18
Conservation, Access and Use in a Museum of Living Cultures
Marian A. Kaminitz and W. Richard West, Jr with contributions from Jim Enote, Curtis Quam, and Eileen Yatsattie

210 Chapter 19
The Challenge of Installation Art
Glenn Wharton and Harvey Molotch

223 Chapter 20
Contemporary Museums of Contemporary Art
Jill Sterrett

229 Chapter 21
White Walls: Installations, Absence, Iteration and Difference
Tina Fiske

241 Index

Contributors' biographies

Jonathan Ashley-Smith studied chemistry to post-doctoral level. For four years he was a trainee metalwork conservator at the Victoria and Albert Museum. Then, for twenty-five years, he was head of the V&A's conservation department. Thirty years' experience in just one museum has limited his worldview. He first wrote about conservation ethics in 1982 when he was young and his views were rigid and idealistic. He is now older.

Erica Avrami is the Director of Research and Education at the World Monuments Fund. She is also a doctoral candidate in Planning and Public Policy at Rutgers University, where her research focuses on sustainability planning and the politics of preservation. She received her undergraduate and Masters Degrees in architecture and historic preservation, respectively, from Columbia University. Erica Avrami formerly served as a Project Specialist at the Getty Conservation Institute, where she was engaged in work regarding the role of values in heritage conservation, preservation planning, and earthen architecture.

Zuzana Bauerová is an art historian and paintings conservator who graduated from Comenius University (1999) and Academy of Fine Arts and Design in Bratislava (1999). She undertook postgraduate studies at the Academia Istropolitana Nova (2001) and Institute of Art History of the Slovak Academy of Sciences (2001–2004). She was a Project Coordinator at the Ministry of Culture of the Czech Republic from 2004 to 2006, a researcher at the Institute of Contemporary History of the Czech Academy of Sciences from 2006 to 2007 and, more recently, has been working on a PhD at the Masaryk University Brno. She lectures at the Academy of Arts, Architecture and Design in Prague, and at the College for Vocational Studies in Information and Library Sciences, and is Project Director of Cultureplus Ltd, a consultancy for cultural heritage preservation and management.

Isabelle Brajer received her Master's Degree in the conservation of paintings from the Academy of Fine Arts, Department of Conservation in Krakow (Poland) in 1983. Since 1985 she has worked as a conservator of wall paintings at the National Museum of Denmark, and currently holds the title of Senior Research Conservator. She has numerous publications in the field of conservation/restoration of wall paintings, which cover a wide range of topics, including theoretical, historical and practical issues.

Simon Cane is Head of Museum Operations at Birmingham Museums and Art Gallery. He has worked in the conservation sector since 1981, starting as a conservation apprentice at Hampshire County Council Museum Service. Educated at Lincoln College of Art and Design and the University of Southampton, Simon is an advocate for the conservation sector and is passionate about communicating the subject to wider audiences.

Chris Caple (BSc, PhD, ACR, FIIC, FSA) acquired his BSc from Cardiff in Archaeological Conservation (1976–1979), and his PhD at Bradford, technology of manufacture of medieval copper alloy pins (1979–1986). Chris was an objects conservator at York Castle Museum from 1984 to 1988; a lecturer in archaeological science and conservation from 1998 to 1996, and is currently a senior lecturer (from 1996 to the present) running the MA in Conservation of Historic Objects (Archaeology) course. He was also the Director of Cadw-funded excavations at Dryslwyn Castle from 1984 to 1995. Chris Caple is the author of *Conservation Skills: Judgement, Method and Decision Making* (2000, Routledge) and *Objects: Reluctant Witnesses to the Past* (2006, Routledge).

Miriam Clavir received her Master of Art Conservation from Queen's University (Canada), and her Doctorate in Museum Studies from the University of Leicester (UK). As Senior Conservator at the UBC Museum of Anthropology, she focused on contextualizing museum conservation values and practices, and on the changing relationship between aboriginal peoples and museums. She has taught for several universities and published many articles as well as the book *Preserving What is Valued: Museums, Conservation and First Nations* (2002, UBC Press). Her current position: Dr Miriam Clavir, Research Fellow, Museum of Anthropology at the University of British Columbia, Vancouver, Adjunct Lecturer, Graduate Program in Museology, University of Washington, Seattle.

Helen Clifford read History at Cambridge before joining the joint Victoria and Albert/Royal College of Art History of Design Postgraduate Course. Her subsequent PhD on a partnership of eighteenth-century goldsmiths gained much from her practical silversmithing work at the College. She has worked on historical and contemporary silver and the decorative arts, on which she has written, taught and curated extensively. She is currently a Senior Research Fellow in the Department of History at the University of Warwick, and runs the Swaledale Museum in Reeth, North Yorkshire.

Dinah Eastop (MA, FIIC, ACR, FHEA) is a Senior Lecturer at the Textile Conservation Centre, University of Southampton, UK. She is the Founding Director of the AHRC Research Centre for Textile Conservation and Textile Studies (2002–2007). Dinah also works with ICCROM's Collection Unit, most recently for the CollAsia2010 programme. She is interested in the dynamic interplay between the material properties and the social attributes of museum objects and its effects on conservation and curatorial decisions.

Jim Enote has explored to a large degree such varied subjects as cultural pattern languages, Zuni architecture as fluxus art, Japanese art after 1945 and, from

1999 to 2004, indigenous community-based mapping. Besides currently serving as Director of the A:shiwi A:wan Museum and Heritage Center, he is a Senior Advisor for Mountain Cultures at the Mountain Institute. He is now camped out at his work-in-progress home in Zuni.

Tina Fiske is based in the History of Art Department, University of Glasgow. She is currently Research Associate for the National Collecting Scheme Scotland, working with seven museums across Scotland to build their contemporary holdings. Her PhD (University of Glasgow, 2004) presented a study of the acquisition and long-term care of contemporary artworks by British regional collections 1979–2004. She teaches postgraduate courses on the collecting and conserving of contemporary art, and has contributed to *Andy Goldsworthy at Yorkshire Sculpture Park* (Yorkshire, 2007) and Andy Goldsworthy *En Las Entrañas del Arbol* (Madrid, 2007).

Cathleen Hoeniger is Associate Professor of Art History at Queen's University in Canada, and her research focuses on the history of the restoration of Italian Renaissance paintings. Her publications include a book on the renovation of paintings in Tuscany, 1250–1500 (1995), an article on the reception of Correggio's *Loves of Jupiter*, and the chapter 'Restoring Raphael,' for *The Cambridge Companion to Raphael* (2005, Cambridge University Press, Marcia B. Hall, ed). Her book, *The Afterlife of Raphael's Paintings*, is forthcoming with Cambridge University Press.

Jukka Jokilehto was born in Finland, graduated in architecture in Helsinki, DPhil at the University of York, England. He first worked as architect and city planner in Finland. Then, in 1971, he attended the International Architectural Conservation Course at ICCROM in Rome. Here, he was appointed Director of the Architectural Conservation Programme, and Assistant to Director General of ICCROM. Retired in 1998, he was Advisor to ICOMOS on UNESCO World Heritage Convention. Internationally known as a lecturer, he has written on conservation theory and practice, including *A History of Architectural Conservation* (1999, Butterworth-Heinemann).

Marian A. Kaminitz is Head of Conservation at the National Museum of the American Indian, Smithsonian Institution (1991 to the present). She was Assistant Conservator at the American Museum of Natural History's Anthropology Department (1985–1991) and Adjunct Professor of Conservation at New York University's Conservation Center (1988–1998). Her MS degree in the Conservation of Artistic and Historic Works from the University of Delaware/Winterthur Museum Program in Art Conservation was awarded in 1984. Coordinator of ICOM-CC's Ethnographic Working Group (1999–2005), she is currently an assistant coordinator.

Jonathan Kemp has over fifteen years experience as a senior conservator and consultant to both public and private UK collections and on international projects in Spain, the Ukraine, and Iran. Currently, he is a senior sculpture conservator at the Victoria and Albert Museum, London. Since 2001, he has also collaborated on various technological projects in the collective ap/xxxxx (http://www.1010.co.uk) including live coding performances, data processing, speculative conferences, and social software events throughout Europe and the US.

Harvey Molotch is Professor of Sociology and Professor in the Department of Social and Cultural Analysis at New York University. Previously, he was Centennial Professor at the London School of Economics and on faculty at University of California, Santa Barbara. Among his awards is the Lynd Prize for Lifetime Achievement in Urban Studies. His most recent book is *Where Stuff Comes From: How Toasters, Toilets, Cars, Computers and Many Other Things Come to Be as They Are* (2003, Routledge).

Salvador Muñoz Viñas is Professor at the Department of Conservation Studies of the Universidad Politécnica de Valencia, Spain. He holds university degrees in Art History and Fine Arts. Before joining the UPV as a lecturer and researcher, he worked as a conservator at the Historical Library of the Universidad de València. Presently, he teaches paper conservation and conservation ethics, and is head of the paper conservation section of the Instituto de Restauración del Patrimonio of the UPV.

Elizabeth Pye graduated from Edinburgh with an MA in Prehistoric Archaeology, then took the Diploma in Conservation at the then University of London Institute of Archaeology. After working at the British Museum on the Sutton Hoo material, she joined the conservation staff of the Institute of Archaeology (now part of UCL) where she is currently Senior Lecturer and coordinates the MA in Principles of Conservation.

Curtis Quam was born and raised in Zuni, New Mexico and has been working as a Museum Technician for the A:shiwi A:wan Museum and Heritage Center for the past seven years. Curtis continues to live in Zuni with two daughters and fiancée Tara. His role in the museum has allowed Curtis to work with community members building ways through a variety of different media to educate youth and interested people about A:shiwi beliefs and history. He hopes to continue working, learning, and teaching in Zuni.

Jonathan Rée is a freelance philosopher and historian based in Oxford, England. His books include *Proletarian Philosophers*, *Philosophical Tales*, *Heidegger*, and *I See a Voice*.

Marcelle Scott coordinates the Masters degree program in Cultural Materials Conservation at the University of Melbourne. She has an undergraduate degree in Cultural Materials Conservation from the University of Canberra and a postgraduate degree in Archaeology from James Cook University. Marcelle has over twenty years' experience in the conservation profession, was National President of AICCM from 1999–2001, and is the Editor of the peer-reviewed *AICCM Bulletin*. Marcelle's research interests relate to conservation pedagogy. Her current research investigates interdisciplinary teaching and learning, and pedagogical aspects of practice-based learning exchanges.

Catherine Smith (BA(Hons) Archaeology, University of Melbourne; BAppSci Conservation of Cultural Material, University of Canberra) has worked as an objects conservator in Australia and New Zealand. She is currently the Vice-President of New Zealand Conservators of Cultural Material (NZCCM), and a lecturer and doctoral candidate in the Department of Clothing and Textile Sciences, University of Otago, New Zealand. Catherine has research interests in material culture, and the ethics and practice of ethnographic objects conservation.

Nicholas Stanley-Price worked as an archaeologist in Cyprus and the Middle East before being converted to conservation. While on the staff successively at ICCROM and the Getty Conservation Institute, he specialized in archaeological conservation and professional education. At the Institute of Archaeology, University College London, he introduced a new MA programme on site conservation and management, before returning in 2000 to ICCROM as Director-General. Since 2006 he has been based in Rome as an adviser on heritage preservation.

Jill Sterrett is Director of Collections and Conservation at the San Francisco Museum of Modern Art. Trained as a conservator, she is particularly inspired by contemporary art, including all forms of new media, and the philosophical and methodological challenges associated with its long-term care. Jill is interested in the ways that collecting and preserving contemporary art calls into question some of the fundamental assumptions underlying the traditions of fine art stewardship and she is committed to fostering the vital collaborations between artists, curators, technical experts, registrars, and conservators which underpin contemporary art conservation practice.

W. Richard West, Jr. is Founding Director Emeritus of the Smithsonian's National Museum of the American Indian. He was Director from its inception in 1990 until 2007, and oversaw the development of its public programs and facilities located in New York City, Washington, DC, and Suitland, Maryland. Richard West was previously Chair of the American Association of Museums and currently serves as Vice President of the International Council of Museums. He is a Peace Chief of the Southern Cheyenne.

Glenn Wharton is Media Conservator at the Museum of Modern Art, New York. He is also a Research Scholar in Museum Studies and the Conservation Center of the Institute of Fine Arts, New York University. He currently serves as Acting Director for the North American group of INCCA, the International Network for the Conservation of Contemporary Art.

Eileen Yatsattie, a Zuni Pueblo tribal member, began making pottery at 13 in 1973. She collects and processes her own clays, pigments, and plants using traditional methods she learned from Zuni elder potter, Josephine Nahohai. Her traditional pottery work continues that of her great-grandmother and her grandmother and ties to Zuni religion and philosophy. She passes on her knowledge through demonstrations, lectures, and through teaching her pottery techniques.

Acknowledgements

The editors would like to thank the following people for their support and advice: Professor Jonathan Ashley-Smith; Dinah Eastop; colleagues at the Royal Academy of Arts Education Department, the Victoria and Albert Museum Conservation Department, and the Royal College of Art/Victoria and Albert Museum Postgraduate Conservation Programme; Eric Richmond; our editors throughout the duration of this project, Stephani Allison, Lahn Te, and Hannah Shakespeare; and, of course, our contributors without whom this book would not have been possible.

Introduction

Alison Bracker and Alison Richmond

Decision-makers and those who act in the name of conservation do things to irreplaceable works of art and design, archaeological artefacts, buildings, monuments, ruins, and heritage sites on behalf of society. As professionals working in the public interest, they contribute to the social process by which the materials and values associated with objects, buildings and sites are transmitted through time. Their role and methodologies therefore demand keen understanding of material culture, and sensitive negotiation of the interconnection between an artefact or site's materiality and social relationship throughout its history, in order to enable both it and its significance to persist.

At the base of this interconnection lies a multifaceted matrix of values that have changed over time and are open to interpretation. Moreover, that matrix, and the society for which conservators care and treat artefacts, monuments, and sites, perpetually shift and diversify. As one conservator notes, conservation is 'a complex and continual process that involves determining what heritage is, how it is cared for, how it is used, by whom, and for whom.'[1] The field's growing awareness of the implicit socio-cultural responsibility that underpins this process has provoked discourses – particularly over the past twenty years or so – concerned with ethical and principled conservation theories and practices.

Yet the past few decades have also witnessed increasing discomfort within the profession with what appears to be a lack of rigorous self-analysis. Conservator Hanna Jedrzejewska has bemoaned the absence of a methodological analysis of conservation ethics,[2] whilst renowned architectural conservator Frank Matero has observed that the profession has thus far 'avoided a critical examination of our own historical-based and culturally based narratives.'[3] But Nicholas Stanley-Price, Salvador Muñoz Viñas, Miriam Clavir, and Caroline Villers,[4] to name but a few, have published seminal texts on the history and theory of conservation and/or its principles. Why, then, does the profession perceive a scarcity of critical self-evaluation? Perhaps because consensus on an over-arching definition of what conservation is does not exist. Or perhaps because conservation theory has emerged from within specialist practices concerned with varied materials (wood, ceramic, stone, paper, textiles and, recently, foodstuffs and other fugitive materials) and object types (paintings, sculpture, installations, artefacts, books, furniture) from within different contexts (collections, buildings, monuments, sites). The museological origins and evolution of these different practices have

Introduction

led to disparate approaches even within the same museum. Finally, perhaps it is because the early phase of conservation theory development, which focused on fine art, only gradually added a body of literature on archaeological artefacts, decorative arts, and architecture that is not easily available to those outside the conservation profession, or even to those in other conservation disciplines. It would seem that the fragmentary nature of the conservation profession confined critical analysis to conservation specialisms, or outside the field altogether, thereby fuelling the perception of a failure to self-analyse.

Whilst the accuracy of this perception may be open to debate, we admittedly felt it keenly. Several years ago, we shared a coincident recognition that our respective professional negotiations of conservation theory and practice had become constrained by the lack of visibility of philosophical and practical inquiry into conservation ethics. At the same time, our colleague Jonathan Ashley-Smith (formerly Head of the Victoria and Albert Museum Conservation Department) put forward the idea that there was a need for 'a book on ethics,' a notion swiftly seconded by our original editor at Elsevier, Stephani Allison. Having worked together for several years organizing annual departmental 'Ethics Days' at the V&A, at which museum staff, external specialists from within and outside the conservation field, and postgraduate conservation programme students joined together to debate topical themes in conservation ethics, we decided to expand upon those fruitful experiences and seized the opportunity and the immense but exhilarating challenge of advancing conservation ethics discourse.

From the beginning, we were committed to bringing together critical thinking from a variety of fields of practice. The conservation profession's codes of ethics have generally been informed by experiences across the disciplines, reflecting the membership of the professional bodies; however, there is still a firm division between moveable cultural property and built heritage. We wanted to embrace the diverse aspects within and outside conservation, not reinforce divisions, in order to provoke the cross-fertilization of ideas from one sphere to another. Our contributors, who come from the fields of philosophy, sociology, history, art and design history, museology, conservation practice and theory, architecture, and planning and public policy, address a wide range of conservation theories, ethics and principles in ways that encourage the reader to compare and contrast *across* specialist areas. But, while their contributions offer many opportunities for comparison, this is not a textbook, nor is it comprehensive. Instead, the chapters herein invoke and stand alongside the current body of knowledge, complement it, and aim to prompt further debate.

Conservation is currently re-evaluating itself in relation to society and acknowledging both its role in assigning and perpetuating cultural value, and its need for greater dialogue outside of the profession. The chapters herein capture thinking at a time when large fluctuations are happening within conservation theory, including the philosophical shift from scientific objective materials-based conservation to the recognition that conservation is a socially constructed activity with numerous public stakeholders. They therefore offer snapshots of how conservation narratives and ethics are being reconsidered, reinterpreted, and reconfigured in this first decade of the twenty-first century.

But what are the key principles under review at this time, and how have they evolved? Arguably, their genesis may lie in sculptor Antonio Canova's refusal in 1816 to restore the fragmentary sculptures brought from Greece to England by Lord Elgin on the grounds that no-one, not even he, could improve on the style of the original artist, and that their fragmentary state should thus pertain.[5] Canova's stance defied the contemporaneous convention of fully restoring fragmentary antique sculptures; indeed, in that same year, Danish sculptor and collector Bertel Thorvaldsen completely restored the sculptures of the pediment of the Temple of Aphaia at Aegina, including the addition of modern replacement heads, limbs, drapery and armour. Canova's refusal to intervene kindled two of conservation's fundamental principles: the desire to preserve the 'authentic' work of art unsullied by restoration (thereby maintaining the aura of the artist's authorship), and the acceptance of damage incurred since its conception (since physical evidence of the object's history seemingly conveys authenticity). Attempts to uphold simultaneously these two principles expose the contradiction at their core, however, for damage incurred over time may result in an artefact or site that bears only the most tenuous link to its creator's 'hand.' Nevertheless, these principles persisted and informed two further concepts intrinsic to twentieth-century conservation theory: the need to preserve the integrity of the original, and the belief that scientific methodology is the best way to do so.

Conservation's faith in science derived from Enlightenment ideas about objectivity, rationality, epistemology, and material evidence. Furthermore, once scientific developments permitted the identification of original materials, the idea that science could therefore pinpoint the artist's 'original intention' (where integrity and authenticity were presumed to reside) developed, giving added weight to the belief in the conservator's objectivity.[6] But over the past two decades, conferences and publications have probed not only the premise of objectivity, but also other concepts that have guided contemporary conservation and restoration theory and practice, including authenticity, minimal intervention, and the conservator as cultural heritage caretaker. The question posed by the title of the 1994 British Museum conference, 'Restoration – is it Acceptable?,' for example, forced those in attendance to grapple with its tacit counter-question – 'Acceptable to whom?' – thereby triggering recognition of society's stake in cultural heritage, its role in ascribing value to the artefacts and sites comprising that heritage, and the fluctuating nature of value itself.[7] In the decade that followed, the field's presumed objectivity came under fire in noteworthy essays by Miriam Clavir, Dinah Eastop, Salvador Muñoz Viñas, and Caroline Villers,[8] all of whom queried conservation's credence in the ideas of scientific truth and neutrality. Villers in particular

Introduction

revealed conservation's heretofore denied subjectivity, noting, 'In practice, conservators are always "writing" the history of the object, as even a decision to do nothing at all constitutes an interpretation articulated through presentation and display,' and declaring, 'The assumption that a conservation treatment is neutral and does not alter meaning is untenable....'[9]

These recent unveilings of the contradictions and fallacies embedded in some of conservation's key principles have compelled those in the field to revisit and rethink the codes of ethics that their institutions, disciplines, and professional bodies have issued. The following essays evince how conservators and others concerned with the production and consumption of cultural heritage understand, internalise, and respond to the ways in which contemporary developments within and beyond the field of conservation are challenging traditional ethics and practice. Though the chapters are highly varied in their scope, focus, and methodology, they all expose the uncomfortable truth of the impossibility of singular and objective truths within cultural heritage care and management. By tracing the agencies and agendas that once drove, or drive today, the development of principles in conservation and its specialized disciplines (Ashley-Smith; Stanley-Price; Jokilehto), moments in history (Hoeniger), countries and communities (Bauerova; Smith and Scott; Kaminitz and West), and new art media (Fiske; Sterrett; Wharton and Molotch); scrutinizing conservation's aims and whether they can be reconciled with future developments (Caple); unpacking the factors through which cultural value is ascribed at any given time (Clifford; Brajer; Clavir); identifying and interrogating the social constructs, processes, and needs with which conservation must engage (Cane; Pye; Eastop; Avrami); and critically analysing the very precepts of conservation ethics (Kemp; Rée; Muñoz Viñas), the authors wrestle with and offer ways of disentangling the ethical dilemmas confronting those who maintain and sustain cultural heritage for today and tomorrow. The resolutions each author proposes invite widespread consideration and debate, as well as future re-evaluation and re-visitation, not least by the authors themselves. For although conservation principles are changing rapidly, they require time and space for debate to occur. As Mark Twain advised on the subject of principles, '...you hang them up to let them season.'[10]

Notes

1. Samuel Jones and John Holden, *It's a Material World* (London: Demos, 2008) 28.
2. Hanna Jedrzejewska, *Ethics in Conservation* (Stockholm: Kungl Konsthogskolan, 1976). See Jonathan Ashley-Smith, Chapter 2, The basis of conservation ethics, 6–24.
3. M. Cassar, M. Marincola, F. Matero and K. Dardes, "A Lifetime of Learning: A discussion about conservation education," *The GCI Newsletter*, 18(3), 2003, 11.
4. See Nicholas Stanley-Price, M. Kirby Talley Jr, and A. Melucco Vaccaro, eds., *Historical and Philosophical Issues in the Conservation of Cultural Heritage* (Los Angeles: Getty Conservation Institute, 1996); Salvador Muñoz Viñas, "Contemporary theory of

conservation," *Reviews in Conservation,* Volume 3, ed. Noelle Streeton, IIC (2002), 25–34; Miriam Clavir, *Preserving What is Valued: Museums, Conservation and First Nations* (Vancouver: University of British Columbia Press, 2002); Caroline Villers, Post minimal intervention, *The Conservator*, Volume 28 (2004) 3–10.

5. Paul Philippot, "Restoration from the perspective of the humanities," *Historical and Philosophical Issues in Conservation of Cultural Heritage* (Los Angeles: Getty Conservation Institute, 1996) 216–229; William St. Clair, *Lord Elgin & The Marbles: The controversial history of the Parthenon Sculptures* (Oxford University Press, 1998) 149.
6. Steven W. Dykstra, "The artist's intentions and the intentional fallacy in fine arts conservation," *JAIC* 35 (1996) 197–218.
7. *Restoration: Is It Acceptable? British Museum Occasional Paper 99*, ed. A. Oddy (London: The British Museum, 1994).
8. See Miriam Clavir, "The social and historic construction of professional values," *Studies in Conservation*, Volume 43 (1998) 1–8; Dinah Eastop, "Textiles as multiple and competing histories," *Textiles Revealed: Object Lessons in Historic Textile and Costume Research*, ed. Mary Brooks (London: Archetype Publications, 2000) 17–28; Salvador Muñoz Viñas, "Contemporary theory of conservation," *Reviews in Conservation*, Volume 3, ed. Noelle Streeton, IIC (2002) 25–34; Caroline Villers, Post minimal intervention, *The Conservator*, Volume 28 (2004) 3–10.
9. Caroline Villers, "Post minimal intervention," *The Conservator*, Volume 28 (2004), 6.
10. Mark Twain, *Mark Twain's Speeches* (Whitefish, MT: Kessinger Publishing, 2004), 61–62.

1

Auto-Icons

Jonathan Rée

The dilemmas of conservation and restoration are usually discussed in terms of great works of art that have either been allowed to decay, prompting accusations of criminal neglect, or else laboriously restored, occasioning charges of criminal damage. But the same issues can also present themselves in everyday life. Old clothes for example – clothes we have been wearing for years, or cast-offs from parents or friends, or purchases from car-boot sales or charity shops: do we want to turn the cuffs and restore the nap and refresh the colours to make them look shop-new, or would we prefer to leave them as they are, a little more ragged at every outing, but bearing honest testimony to the scenes they have witnessed and the places they have been? And what about thinning hair or roughened voices, stretch-marks, lopsided smiles or chipped, yellowing teeth: are they to be prized as trophies from our voyages round life's extremities, or discretely concealed as unseemly intimations of decrepitude and imminent collapse?

The same questions arise with buildings. You only have to look at photographs of towns and cities a hundred years ago to lament the haste with which perfectly serviceable old shops, houses and apartment blocks, or entire streets and neighbourhoods, have been knocked down to make way for brash modern structures which, to our eyes at least, have aged very badly indeed. Consider too the removal of wainscoting, fireplaces or glazing bars, the installation of heating radiators or indoor toilets, the prodigal kitchen extensions and loft and basement conversions that characterize the houses where most of us spend the private and domestic parts of our lives.

Long before the housing bonanzas of the twentieth century, the problem was articulated by the utopian socialist William Morris, who founded the Society for the Protection of Ancient Buildings in 1877, with active support from the utopian Tory John Ruskin. The argument of the Society can be summed up in Morris's phrase, 'anti-scrape.' Buildings, for Morris, were like people or animals or plants, and it was in their nature to grow feeble and bowed and crotchety with age. Damage, injury, corrosion and decay should be avoided if possible, but if repairs became necessary they should take the form of minimal protection rather than active restoration – what Morris called 'daily care,' aimed at patching and mending with 'no pretence of other art.'[1] If the walls of old buildings in Oxford are pitted and crumbly with age, or the stone steps worn down with use, then that is part of their charm and their dignity, and a zealous mason who scraped away the work of time and laid on a veneer of fresh-cut stone would be perpetrating a kind of insult if not a blasphemy. Restoration, for the anti-scrape party, was destruction and desecration by stealth.

Architecture, as Ruskin had written nearly thirty years before, is an art of memory: a building is not just an object of present experience but a memorial and a monument, a relic of the past and a communication to the future. Old age is the test of a good building, and there is, he said, 'an actual beauty in the marks of it:'[2] architecture, for Ruskin, possesses a 'light, and colour, and preciousness' that become manifest only in a 'golden stain of time.'[3]

> Do not let us talk then of restoration. The thing is a Lie from beginning to end. . . . The principle of modern times . . . is to neglect buildings first, and restore them afterwards. Take proper care of your monuments, and you will not need to restore them. . . Watch an old building with anxious care; guard it as best you may . . . bind it together with iron when it loosens; stay it with timber when it declines; do not care about the unsightliness of the aid: better a crutch than a lost limb; and do this tenderly, and gently, and continually, and many a generation will still be born and pass away beneath its shadow. Its evil day must come at last; but let it come declaredly and openly, and let no dishonouring and false substitute deprive it of the funeral offices of memory.[4]

Looking back on these words in 1880, Ruskin declared them 'more wasted'[5] than anything else he ever wrote; and even if you resist his piteous sentimentality, you may find it impossible not to sympathize with his feeling that buildings can have a venerable life of their own, or even a personality – that they will flourish when granted the freedom to be their own idiosyncratic selves, and that when the time comes it may be better to let them die with dignity than turning them into ghastly simulacra of their former glories, or death masks or mummified corpses in an architectural mausoleum.

It is easy enough to contemplate the mortality of ordinary houses, however lovely: we accept that at some point they will come to the end of their useful life and be reduced to rubble; but even then we might be comforted to know that a few specimens of a typical two-up two-down cottage, a Victorian villa, or a suburban front room have been frozen in time and reconstituted inside a local museum. And when it comes to unique public structures like the Radcliffe Camera or Rouen Cathedral – beautiful in themselves, and invested with love and significance over many generations – it is hard not to feel that they ought to be preserved at any cost, and in their original setting, as long as civilization survives. The church restorations that angered Ruskin and Morris may have involved fudges or even lies, but they possess their attractions too, and twenty-first century cities are looking better than they might thanks to the fad for refurbishing old buildings, however artificially, rather than knocking them down and starting again.

The anti-scrape faction had a good point when they denounced the practices that threatened to turn living buildings into deathly waxworks; but once the possibilities of patch-and-mend have been exhausted, the option of thorough restoration will start to look more attractive. Any building that is blessed with some kind of world-historical singularity has a certain claim on our care, and none of us would want to be the one to call in the wrecking-ball, even if we risked putting ourselves on the wrong side of the Morris–Ruskin line. Anti-scrape purism will begin to look like pig-headed self-indulgence if it means depriving future generations of objects they might regard as a legitimate inheritance.

In an essay written in the last months of his life, the Utilitarian philosopher Jeremy Bentham suggested that when people die, their bodies should be chemically preserved as permanent monuments to their own past existence. He looked forward to a future in which these embalmed corpses – 'Auto-Icons' as he called them – would be displayed in public buildings, providing unimpeachable likenesses of the departed at very low cost (no artists' fees payable), and in accordance with the best democratic principles ('every man would be his own monument,'[6] as he put it).

Bentham's notion of the Auto-Icon deserves to be extended to cover any objects that are removed from ordinary cycles of maturation and decay in order to become permanent representatives of their former transient selves – including, notably, the contents of museums and historical libraries and galleries. To consign an object to a museum is to designate it as an Auto-Icon, and thus to generate a series of painful dilemmas over access and preservation: there is after all no point in keeping your collections immaculate if you do not let anyone consult them, but you cannot do that without exposing them to conditions that are liable to damage them and curtail their lease on life. But an anti-scrape policy is not an option: you cannot banish restoration as a lie, and embrace decay and dilapidation in the name of honesty and truth, since the whole point of taking objects from somewhere else and putting them in museums is to shield them from the ordinary attrition of old age.

Hence the repeated resort to the criterion of the 'original state:' objects in museums should be maintained in their original state, or if necessary restored to it, since that is what they are there for. The principle is not a bad one, but not especially helpful either – partly because of certain well-rehearsed theoretical difficulties about the nature of authentic originality,[7] but mainly because of pressing practical problems in knowing what the object would have been like at any given stage in its career, and deciding when it would have been most fully and originally itself (you cannot assume that things are always at their best when brand-new). William Morris noted that those who profess to restore a building to its pristine condition inevitably operate within the perspective of their own time: they perpetuate oversights and insensitivities that will not bother most of their contemporaries though they will become painfully obvious to later generations.[8] The restoration of St Alban's Cathedral, for example, appeared impeccably gothic when it was carried out in the 1860s, but to later eyes it looks deeply Victorian; and touched-up paintings from the Italian renaissance are liable to look as if they issued from the studios of the Pre-Raphaelite Brotherhood.

The problem of obsolescence and restoration in museums has an analogue in the field of translation. The language of a classic original text – the words of Tolstoy, Shakespeare or Cervantes for example – will come to seem archaic as time goes by, and perhaps challenging or awe-inspiring too, but they will hardly stand in need of updating: readers of Russian, English or Spanish will be prepared to grapple with the difficulties of their ancient classics and on the whole they will be well rewarded for their pains. The language of a translation, however, will soon grow obsolete, and new translations will become necessary: original texts do not grow old as those that translate them grow old.

But the burdens of responsibility that weigh on translators are negligible compared with those imposed on the conservators, curators and restorers. Translators, like editors, deal with texts rather than physical objects, let alone icons or Auto-Icons; their materials are sequences of symbols that can be copied over and over again without any loss of significance – and they can do their worst with them without putting the original at risk in any way; whereas retouching a painting or replacing a sculpture's lost hands may endanger its very quality as a uniquely precious work of art.

Technology helps: mechanical reproduction of paintings and sculpture means that the originals can be left exactly as they were, as immaculate as Auto-Icons can be, offering the smallest possible offence to the most fanatical fetishists of original works of art. Digitally engineered doubles can be dismantled, altered and restored in as many styles and tastes as we like, just as literary texts are translated and retranslated, and each new version will win a welcome as long as it reveals something new, and refrains from laying claim to absolute authenticity. It is often said that every act of translation, however faithful, is an act of betrayal as well, and the same applies to the conservation and restoration of works of art; but if there is something sombre about it, there is also some humble comfort in the reflection that everything we try will fail in some way, and nothing can ever be perfect.

Notes

1. William Morris, "Manifesto of the Society for the Protection of Ancient Buildings on its Foundation in 1877," *William Morris: Artist Writer Socialist*, ed. May Morris, vol. 1 (Oxford: Basil Blackwell, 1936) 113.

2. John Ruskin, "The Lamp of Memory," *The Seven Lamps of Architecture (1849): The Works of John Ruskin*, eds. E.T. Cook and Alexander Wedderburn, vol. 8, 39 vols (London: George Allen, 1903): 221–247, 235.
3. Ruskin, p. 234.
4. Ruskin, p. 234.
5. Ruskin, p. 245n.
6. Jeremy Bentham, *Auto-Icon; or, Farther Uses of the Dead to the Living: A fragment* (London: Privately Published, 1842) 3–4.
7. *See for example* William K. Wimsatt and Monroe C. Beardsley, "The Intentional Fallacy," *Sewanee Review* 54 (1946): 468–488, revised and reprinted in his *The Verbal Icon: Studies in the Meaning of Poetry* (Lexington: University Press of Kentucky, 1954): 3–18. Roland Barthes, "La mort de l'auteur (1968)," translated by Stephen Heath and Roland Barthes, *Image-Music-Text* (London: Fontana, 1977) 142–148.
8. 'Those who make the changes wrought in our day under the name of Restoration, while professing to bring back a building to the best time of its history, have no guide but each his own individual whim to point out to them what is admirable and what contemptible.' Morris, p. 110.

2

The Basis of Conservation Ethics

Jonathan Ashley-Smith

Introduction

This contribution is about past and present influences on the principles that guide conservators in their work. The title implies that there is some discrete and recognizable foundation for the belief system that directs the everyday activities of conservators. The discussion that follows examines the limits of such an assumption.

The way conservators behave changes with time and location, as does the way they feel about the behaviour of their peers. Even at one single time and place, it is not certain that two conservators would be in detailed agreement about a proposed treatment. The things that conservators choose to worry about are influenced by the changing social and political environment outside their workplace. The way that conservators work is heavily influenced by the internal politics and pressures of their employment. A single historic basis for conservation principles seems unlikely. Some past publications or events may not be as influential as might be predicted. In many walks of life theory follows practice, rather than the other way round. It is possible to develop all the necessary protocols and practical skills without ever learning the theory.

There are two distinct sections to this chapter. The first, which deals with the prehistory of conservation, is highly speculative and only superficially academic. The second section, dealing with recent history, relies to some extent on my own observations of conservators at work, and is necessarily subjective. It covers a period during which my own beliefs about what is acceptable behaviour within conservation were changing. My documentation of historic change may be merely the changing views of an evolving observer.

The changes in conservator behaviour that I have observed over the last thirty-four years can be summarized as increasing involvement with management of collections and projects, and decreasing physical interaction with individual objects. The availability of education and training has increased; job security in institutions has decreased. Conservator organization has increased and become more political. Concepts such as reversibility and minimal intervention have recently been re-examined by individuals, but have not been universally abandoned as ideals.[1]

The prehistory of conservation: the easy history

It has become conventional to trace the history of conservation/restoration practice from a beginning somewhere during the Italian Renaissance when artists were rediscovering the practices and products of classical sculptors. The conventional history of conservation theory takes in John Ruskin and Eugène Viollet le Duc, and climaxes with the work of Cesare Brandi or Salvador Muñoz Viñas (depending on your age and nationality).[2] The development of an overview for conservation practice is traced through a series of documents and declarations, the most critical arguably being Venice (1964), Burra (1988,1999), Nara (1994) and Yamato (2004).[3] Most of such charters concern monuments and sites, and should be interpreted with care when considering museum objects. The history of what we think of as a defined conservation profession takes recognizable shape sometime during the first half of the twentieth century.[4] The attempts at self-regulation by conservation professionals are traced through a series of codes starting with the Murray Pease report at the start of the 1960s, and ending with a number of superficially similar documents adopted by groups of conservation professionals in different countries world-wide, e.g. the United States, Canada, and Europe.[5] This series of published texts – books, charters and codes – all dated and readily available to scholars, provides an academically crisp history that charts the emergence and acceptance of new ideas. But it would be simplistic to think that these textual milestones actually represent the beginning of some new phase or the end of an era. There is a less well-recorded history leading up to each publication. And there may be many stories that do not result in any record, if only because the relevant knowledge and the associated mode of behaviour are assumed to be universally accepted, too obvious to merit discussion.

It would be unusual for someone to suddenly conceive totally new ideas for better ways of behaving, to publish them, and then sit back and wait to see society change (although the Ten Commandments might be an exception).[6] When Emily Post wrote *Etiquette in Society, in Business, in Politics and at Home* in 1922 she did not suddenly conceive a whole new way for people to behave in polite society.[7] There was already a set of behaviours that, although continuously evolving as a result of social and economic drivers, was deemed to be acceptable to a certain group at a certain time. Publication was an attempt to stop evolution and to encourage the untutored to sign up to this fossilized way of behaving. Judging by behaviour at Ashley-Smith mealtimes, the attempt failed. But any attempt to write a set of rules with universal and eternal validity is bound to fail (the Ten Commandments being no exception).

Ethical concepts exist before the professor begins to teach them or before the influential book is published. Ethical principles exist before groups of like-minded people affirm their like-mindedness in declarations and codes of practice. So where do these concepts and principles come from?

It is worth noting at this point that the majority of the rules in each of the published codes for ethical behaviour relate to human–human interactions. Relatively few of them relate to human–object or human–collection interventions. Even James Beck's 'Bill of Rights' for works of art, written for the protection of individual objects, is more about the organization of arts institutions than guidance on good restoration.[8] The following passage from the introduction to the UKIC *Code of Ethics and Rules of Practice* gives a good illustration of the relative importance of people and objects:

> This code is based on the question of what makes the profession creditable and respectable? The main answers are: honesty in dealings with clients, employers, employees and colleagues; giving good and fair advice; being aware of one's limitations; carrying out conservation work to the highest possible standards and not damaging objects; charging fairly for work. From this is (sic) can be seen that a code of practice can be distilled down to treating all persons equally, honestly and pleasantly; maintaining the utmost respect for the objects, whatever their value or rarity, and striving to increase knowledge and understanding of the profession.[9]

Yet it is the physical interaction with heritage objects that distinguishes conservation from other heritage activities. Human–human interactions are generic activities requiring easily transferable skills. It is the human–object interaction that will be at the centre of the following arguments.

The prehistory of conservation: gene-culture co-evolution

I have used the term evolution to describe the constant change of acceptable behaviour brought about by changes in the social environment. Perhaps some aspects of attitudes to conservation and restoration can be explained in terms of human evolution. The battle for supremacy between nature and nurture in the determination of human behaviour is still vigorously fought. Although the inheritance of acquired characteristics is disputed, there is some agreement that in the very early development of humankind there was co-evolution of genes and culture.[10] Up until the Neolithic period, the development of cultural characteristics was taking place at the same slow pace as genetic evolution. After that, and especially with the changes brought about by 'civilization,' cultural evolution was too rapid to be accommodated in genetic change. This means that the modern human must live with inherited propensities and reactions that relate to the survival of a plains-dwelling hunter-gatherer. Some of these, such as a fight-or-flight reaction on meeting strangers, may be somewhat of an embarrassment to a modern city dweller.

There may be types of behaviour that would eventually be reflected in current principles of conservation that favoured the survival of our Stone Age forebears. Conservation, in the sense of not throwing useful things away, seems to be a behaviour that would favour survival. If it uses up time and calories to put an edge on a flint tool, it is advantageous to hang on to it and keep using it until it is no longer serviceable. Conservation, in the sense of continuing maintenance, would seem to be an advantage for survival. You will starve if you have to mend your hunting equipment at the moment you spot your prey. It does not require a giant leap of imagination to suppose that restoration, in the sense of replacing parts of a useful object with new parts made from similar materials, could well be a hard-wired inherited trait. There is a limit to the size and weight of the burden you can continue to carry round with you, moving from one source of food to the next. Hauling around a collection of unrestored tools, even though respecting their authenticity and historicity, seems unlikely to confer any advantage. The restorer survives to continue the species, while the minimal interventionist staggers under the weight or starves to death.

In 1945, American anthropologist George P. Murdock drew up a list of 67 cultural universals.[11] This is a list of social behaviours common to the hundreds of societies that had been studied to that date. The list of human universals was updated in 1991, and its genetic origin has been convincingly argued by Edward O. Wilson.[12] Nearly all of these behaviours relate to what would now be termed the 'intangible.' It includes things like courtship, dancing, divination, language, rituals and taboos. Only a handful, such as decorative art, housing, tool-making and weaving, relate to tangible and possibly lasting objects. Only one universal, rules of inheritance, implies a notion of keeping physical things for future generations. It seems unlikely that specific social manners relating to deliberate yet subtle changes in the nature and appearance of tangible material could be shown to be innate human behaviour.

It has been proposed that the concept of contractual agreement is universal to all cultures, which suggests genetic heritability.[13] Modern behaviours, such as collecting and storing objects for future generations, imply a contractual agreement, albeit with persons unknown. Allowing decay through neglect or overuse can be seen as a violation of that contract. Preserving original values through restoration is one way of keeping the promise. But if modern sensibilities interpret the restoration as faking, this awakens a historic innate disgust at the breaking of the agreement.

The Prehistory of conservation: gene–meme co-evolution

In the way that genes can be thought of as working selfishly for their own survival and replication rather than for the benefit of the social organism carrying them, it has been argued that memes can act in the replication and transmission of behaviour. A meme is a unit of cultural transmission or, according to Susan Blackmore, a unit of imitation.[14] Richard Dawkins cites 'tunes, ideas, catch-phrases, clothes fashions, ways of making pots or of building arches' as examples of memes.[15] Religions are seen as groups of memes with high survival value.

At the very minimum, genetic evolution has given all humans similarly programmed sense organs and brains, giving them a developmental bias called 'prepared learning.'[16] This means that humans are innately prepared to learn certain behaviours and predisposed to avoid others. Humans are innately set up to see and hear things and then to set about imitating them. Thus useful things like multiplication tables and less useful things like the crazy frog ring-tone are replicated and spread. Humans are the physical hosts needed for memes to spread. The meme does not need to be useful to the host. It does not even need to make sense or be beneficial. As an example, Blackmore argues that it was through infectious imitation, rather than an understanding of its long-term benefit, that agriculture became fashionable. Farming uses far more energy and time than hunting and gathering. Being tied to one location makes the farmer more vulnerable to drought, flood, disease or attack. So the farmer seems to have chosen a risky life of endless toil for no obvious benefit.

The meme–gene comparison has many critics, but it is only an analogy, and analogies can always be pushed so far that they fail. The importance of the 'meme as replicator' idea is that it provides a plausible mind-model that explains what we actually observe. People do not always appear to act in a way that we, the sensible few, think of as logical and correct. It explains why ways of thinking and behaving can spread and survive without the 'hosts' being aware of their role in the diffusion. Using this argument, it would not be surprising to find that some conservation principles had spread and become entrenched without being of obvious benefit either to the objects or the conservators. The rise of the scientific paradigm in conservation and the denigration of restoration might be two examples of aberrant and over-reaching fashions.[17]

Springs of action

Memes spread without motivation, but it is tempting to believe that human decisions, to do one thing and avoid another, need some motivating factors. In his *Springs of Action* Jeremy Bentham suggests pleasure as the motivator for certain forms of conduct, and pain as leading to aversion to other options for conducting one's life.[18] This idea can be brought up to date and made more personal by asking about our everyday lives: 'What makes me feel good?' and 'What makes me feel bad?.' Although this approach seems simplistic, it is a good way of finding what attitudes are innate or subconsciously entrenched without immediately trying to solve specific ethical problems. It has the advantage that it answers difficulties raised by other approaches where problems with altruism arise. Sacrificing my wealth or health just 'makes me feel good.'

Here are two suggested lists:

Feel Good	Feel Bad
action	change
discovery	loss
success	failure
integrity	deception
familiarity	neglect
tranquillity	wantonness
involvement	exclusion
validation	criticism

They indicate activities or events that lead to lasting levels of feeling good or bad, rather than the transitory feelings arising from a good meal, alcoholic intoxication or sex. These are not meant to be exhaustive lists, but looking for new words usually results in finding close synonyms rather than whole new areas of feeling. For instance 'safety' and 'conformity' are close relations to 'familiarity.' Nor do the two lists form a series of dichotomous pairs; there are opposites even within a single list. Both action and tranquillity can make you feel good, but not usually at the same time. Similarly, the same thing can make you feel good sometimes and bad at others. Thus action, doing something you find enjoyable, may well lead to change. But change can be very unsettling.

Most of the sixteen words above should be familiar, and though each can have a range of meaning, their location in one of these two lists should be uncontentious. One that may be unfamiliar is 'wantonness.' This relates to behaviour by others that appear to lack motive. So graffiti on buildings and public sculpture seems like wanton (motiveless) damage. It upsets me, but for the 'vandal' there may be both motivation and pleasure.

If you allow that doing something that makes you feel good is good behaviour, and doing or experiencing something that makes you feel bad involves bad behaviour by you or someone else, you have a personal set of ethical guidelines. Relativist chaos would ensue if everyone's personal set of ethics were unique. Chaos is avoided by invoking the positive drivers 'involvement' and 'validation' and the negative constraints of 'exclusion' and 'criticism.' Individuals are encouraged to sign up to group morality.

Codes of conservation ethics, like the 'shalt nots' of the Ten Commandments, attempt to control certain individual feel-good behaviours. Even though the individual may feel better, there will be others, possibly a large number of people, who will feel bad. Conservation–restoration treatments should make you feel good because they require 'action,' they may lead to 'discovery,' will rely on 'involvement' with clients and specialists and end in the 'success' of overcoming problems. These are all good outcomes but they will not result in 'validation' if others feel the treatment has caused 'change,' 'loss' or 'deception.' Well-intentioned treatment may include 'change' in stability, 'loss' of dirt and the 'deception' of an inconspicuous neutral infill. The distinction between what is approved and what deemed unacceptable is quite subtle.

There is an important quantitative element in this ambiguous balance induced by ethical guidelines. Don't do *so much* of the feel good stuff that it makes others feel *more than a little bad.*

It is certainly interesting, and may be relevant, that the 'good' list contains factors that tend to make people underestimate personal risk, leading to actions that we associate with confidence and courage. The bad list contains factors that make people overestimate risk, and consequently worry too much. It has been my observation that the greatest applause at conservation conferences goes to people who have dared to do something, rather than to those whose risk aversion dictated studied inactivity.

The end of prehistory

There are a number of possible points of view about the source of ethical principles. The origins may be:

> transcendental
> intellectual
> fashionable
> traditional
> visceral
> genetic

This is a top down list, with the living gods at the top and the 'blind watchmaker' at the bottom.[19] The first part of this chapter concentrated on the bottom of the list with genetic and memetic origins that lead to the unthinking visceral gut reaction that something is right or wrong.

The transcendental explanation, that ethical guidelines come from somewhere beyond human experience or knowledge, is unlikely, but it is probably not disprovable. Guidance for conservation practice does not appear to come from Mt Olympus or Heaven. The gods of the ancient Greeks and Romans were not given to ethical behaviour, let alone ethical guidance. The choice of right or wrong ways to live was very much left to humans.[20] Religions such as Buddhism, Islam and Christianity provide more specific guidance on lifestyle through the teachings of earthbound messengers. The major writings are now available in several English translations and fully searchable on the Internet, so it's fairly easy to check for the presence of words, phrases or ideas.[21] Some of the stories, such as the great flood in the Old Testament, provide illustrations of nature conservation, the precautionary principle and risk management, but it is difficult to find any useful advice on how to treat physical objects of great value or significance. The main reason for this is that mankind is asked to transcend the need for possessions. We are advised to avoid attachment and not to store up treasures on earth. For the peoples among whom the earthbound prophets roamed, physical wealth was mostly four-legged. Early concepts of wealth would be based on sustainability of the flock (or herd) rather than the long-term conservation of any individual object.

The intellectual approach to determining ethical behaviour may work for recording or reviewing differences in declarations of principle, but I have come to reject the view that ethical concepts are created just by sitting and thinking about them. Thirty years ago I was sufficiently arrogant to believe that intellectual genesis was possible. This is exemplified by my dogmatic interpretation of the first United Kingdom Institute of Conservation Code of Ethics, written less than 10 years after I first began learning about conservation.[22]

Tradition and fashion in conservation ethics: what I observed

I started my immersion in the conservation environment on the first day of January 1973. Over the next four years, I was taught conservation skills by two craftsmen who had received no conservation training. In the years before I joined the metalwork conservation studio, they had never attended any national or international conservation conference. They did not join either IIC or its UK group, which would have been the two most obvious choices for external professional involvement at that time. In the four years I worked in the studio I never saw them open a conservation journal. In the twenty or so years that I knew them (first as their apprentice and then as their boss) they never went to a lecture or read a journal. These two men took their coffee breaks at their benches. Later, when that was deemed risky on health and safety grounds, they retreated to a small office. They never joined in any of the groups of conservators, scientists or curators that gathered elsewhere at break time. Visits to the studio by curators were rare. Visits by the Head of Department were rarer. Yet I learned some fundamental ethical principles from these craftsmen. So how did my mentors learn to work ethically?

Some traditional ways of working can, in retrospect, be interpreted as following currently fashionable ethical concerns. Of my tutors, one craftsman had basic training in silversmithing and the other had started work in the museum blacksmith's shop. I learned how to deform metal without breaking it. Once the skill is acquired this is, to the unaided eye, a perfectly reversible process. If you really wanted to, you could remove a dent from a silver vessel and then put it back again, over and over again.

I learned how to join metal to metal. With sterling silver and hard silver solder this is a reversible process. You can make and unmake the join at will a large number of times. This level of retreatability is also feasible with copper alloys and lead–tin solders. A less elegant sort of reversibility is possible with the poor craftsman's solder, epoxy resin, although unlike the solder, the resin is irreversibly lost at each re-treatment.

I learned that if a silver object has been repaired with lead–tin solder, it can never again be repaired using silver solder. At the higher temperature needed to melt the silver solder, even small traces of lead can cause extensive damage, 'eating' away the original silver object. By studying traditional craft knowledge I learned to distinguish what was safe (good) behaviour and what was damaging (bad) behaviour. Good practice was independent of conservation theory.

The lacquers that I was taught to use to protect clean metal surfaces had been in use at the British Museum since the 1950s, but they had not been developed specifically for museum conservation use. They were the products of industrial development. The predicted service life of the lacquer layer was short compared to that of the underlying metal component, so there was an automatic expectation by the manufacturer that periodic re-treatment would be necessary. Using technology from outside the museum environment, I was given a lesson about the desirability of reversibility.

The restoration techniques I used were safe when applied to objects from the Victoria and Albert Museum's (V&A) metalwork collections, which mostly date from after 1000 AD and have never been buried. Techniques that are closely related to the mechanisms of manufacture can be used when the object is in a physical and chemical state that approximates to the state at the time of creation.

Most objects were complete or nearly so. Where bits were missing it was possible to make replacement parts using techniques that were similar if not identical to the mode of manufacture. So I learned how to find evidence for the nature and appearance of the missing pieces and to make inconspicuous yet identifiable replacements, behaviour that is fully allowable under modern ethical guidance. Marking the replacement parts of an object to identify the craftsman responsible for the most recent intervention is an old tradition continued by my teachers. It is a continuation of the traditions of the craftsmen who originated the objects.

The material dictates the way the object decays or becomes damaged. The material dictates the degree to which manufacturing methods can be used in the conservation treatment, which can effectively be thought of as the first re-treatment. The nature of the material even dictates how easy it is to mark a replacement piece as non-original. If attitudes governing conservation behaviour can come directly from the nature of the material, or from industrial use of similar materials, it follows that certain behaviours may be specific to each particular specialist discipline. It also follows that behaviours interpreted in retrospect as ethical, and therefore fitting universal guidelines, may well have developed independently and without external influence within specific trades and disciplines. Even within one institution such as the V&A it is possible to see distinctly different attitudes in the different specialist sections.[23] The variation is even more obvious when comparing different types of institution.

The similarities between the conservation of decorative arts objects and the conservation of archaeological remains may be only superficial likenesses invoked to prove the unity of the conservation profession. Burial can induce large physical and chemical changes in objects that distance them from their original manufactured state. New techniques have to be developed for what are essentially new materials. The development of specialists to handle these new materials does not require an intimate understanding of, or craft sensitivity to, the physical properties of the old materials from which the archaeological object was created. This alone is enough to suggest that archaeological conservation ethics derive from a different starting point to those that developed in the decorative arts. There are several other reasons why the different ethical concerns started out separate and can never completely converge. Archaeology generates immense quantities of incomplete objects, some of which are extremely unstable once recovered from the ground or the sea. Much of the material derives all of its value from juxtaposition to other objects and the context of the find. By contrast, even a large decorative arts museum will have a much smaller number of objects. These will be more complete, much more stable, and will have already passed through an intrinsic value filter on acquisition.

Similar arguments could be made to explain the differing attitude to treatment and care found in the conservation of architecture, archives and natural history collections.

I came into conservation as a scientist and soon became aware that some treatments can destroy evidence that might be elicited at a future time through scientific examination. However, the possibilities for information retrieval are again dependent on material and on collection type. It would be an extreme precautionary action to preclude any form of treatment for all objects. As with all conservation decisions, it is a balance between known current needs and possible future needs.

Tradition and fashion in conservation ethics: what I heard

When I started out in conservation, my colleagues, regardless of their specialization, had two stock slogans to help with their decision-making: 'every case must be judged on its own merits' and 'it's part of the history of the object.'

Both slogans are liberating rather than constraining. They do not give instructions about what must be done nor dictate what should never be done. Although easily interpreted as incitements to laziness they are actually encouragements to think before acting. The lazy interpretation is that a conservator does not need to learn, as no amount of previous knowledge can help. Since every damage and every accretion is part of the object's history, they must not be removed and so it would be wrong to start any work.

A more positive interpretation of the history argument is that every object has more than one story to reveal. Removal of dents and dirt may be taking away the possibility of telling some of the stories, even though restoration would make the telling of the 'main story' much easier. A positive outcome of every case being different is that standard treatments are not applied as part of an unthinking routine. So both slogans provide very good ethical guidance.

When I first heard them, both expressions were spoken as if in inverted commas and with a certain mock solemnity. These days the solemnity remains but any sense of humour is long gone. A quick Internet search shows that 'part of the history of the object' is still current in conservation discussions. But it has evolved and taken on a range of connotations that would have been unusual in the early 1970s. Collections management considerations, such as object location, accession documentation and treatment records, are now all part of the history. In the Canadian code of ethics the expression 'part of the history' is used only about reports and documentation and not about the 'cultural property' itself.[24]

Because they are always employed with a minimum of variation in the words used, they appear to be quotations of readymade maxims. Yet they do not appear in discussions on the limits of restoration in any of the early charters or codes of practice on which current conservation philosophy is supposedly based. They do not appear in the classic writings on conservation practice or conservation theory. It seems unlikely that the idea that every case is different could be the basis of a universal theory. Indeed Joyce Plesters used it as proof that there could not be just one theory of conservation.[25] Not surprisingly, 'every case must be judged on its own merits' was not invented within conservation but is a long-accepted principle in law. Even there it carries some ambiguity, as law also requires the accumulation of legal precedent.

The observation that everything that happens becomes a part of history is a truism and obviously not limited to conservation studies. So it seems that two bits of guidance that steered my early thinking about conservation ethics, and are still used today and still valid, are examples of what Muñoz Viñas calls 'the revolution of common sense.'[26] They allow the consideration of options; all that is missing is the moralistic insistence that some options are to be condemned.

Tradition and fashion in conservation ethics: what I read (or failed to read)

Ethical guidance may be generated spontaneously from craft understanding of materials and is certainly somehow appropriated from other areas of human activity. But the answer to the question 'how do ethical memes spread in the conservation arena?' has to be 'not very fast, not very far and not very evenly.' A search of the Internet, library catalogues and conservation databases for 'first mentions' supports the idea that some individuals within the profession are very quick to appropriate ideas from other disciplines. However, the diffusion into the mainstream of conservation is very slow.

Until 1972, I had been studying chemistry. I then looked for a museum job. As preparation for my first job interview I read *The Conservation of Antiquities and Works of Art: Treatment, Repair and Restoration* by Plenderleith and Werner, for a long while the definitive version of conservation as practiced in the UK.[27] Unfortunately, it is a recipe book and contains no ethical guidance. I joined IIC in 1973 and consequently had *Studies in Conservation* delivered regularly to my door. No article has ever appeared in *Studies* with the word 'ethics' in the title. Only 17 of 1920 papers in the searchable index of IIC publications have 'ethics' as a keyword. The first of these is from 1975, a quarter of a century after the foundation of IIC.[28] Since most of those 17 instances are conference papers, it seems that the place to learn about conservation ethics, outside of the charters and standards of conduct, is at a conference.

Conservation conferences consist, for the most part, of people seeking validation by preaching to the converted. Cruel though this may sound, the assertion is backed up by the verifiably limited range of the disciplines of the authors and the low attendance of congress participants from different areas of heritage activity. Architects, curators, librarians, archivists and registrars all hold their own conferences. The format of ICOM-CC triennial conferences allows small sub-categories of conservator to speak only to those with a narrow common interest. Since only 15 of more than 800 IIC conference papers have 'ethics' as a keyword, it seems that the majority of the memes to which conservators are exposed concern practical activities, whether preventive or interventive. So just as at work, the main influences are the task and the work environment, rather than any high-level discussion about the meaning or morality of the work.

Last year was widely celebrated as the centenary of the birth of Cesare Brandi. His book *Teoria del Restauro*, published in 1963, is currently being hailed as the most influential publication of the twentieth century. But although the man himself may have been influential at an international level in subtle and political ways through his work with UNESCO, it is hard to trace the influence of his book in my own ethical development. I was not aware of its existence until after I had written several papers on ethics. I did not even become aware of the name when extracts first appeared in English in 1996. Back in 1981 I was part of a group of five conservators who wrote the first draft of a guidance for practice for the United Kingdom Institute for Conservation. We were all fairly young and fairly British; none of us was a fine arts conservator. We drew our inspiration from the American Institute's document, which was itself based on the Murray Pease Report. The Murray Pease Report was the work of a committee convened in 1961; the final draft was approved at around the time that Brandi's masterwork was published. It makes no mention of Brandi and predates the Venice Charter, another document that is seen in retrospect to have been hugely influential on conservation practice.

I am not the only person who remained so ignorant. Hanna Jedrzejewska, eminent international conservator, speaking in 1975, says that 'no methodological analysis has been made' on the subject of the ethics of conservation. Her own analysis of the subject was deemed sufficiently important for her lecture to be published as a booklet in the following year.[29] Her references are English or Polish, not Italian. If you look at papers from the conferences organized by the British Museum in 1994 and 1999, which were deliberately targeted at ethical issues, you will find Brandi cited mostly by Italian authors, rarely by fine arts conservators and never by decorative arts conservators.[30] English-speaking conservators are more likely to refer to English texts, e.g. Barbara Appelbaum on reversibility and retreatability. Appelbaum herself does not cite Brandi.[31]

Ethical ideas become channelled or blocked by the barriers of language and the gulfs between disciplines.

The continuity of ideas

As noted in the introduction, the development of ideas does not happen in discrete steps, each prefaced by a salient publication. It is always possible to find some record of earlier discussions or actions, dealing with some aspect of the problem, that precede the milestone text. For instance, Michael von der Golz writes about discussions on restoration published in German in 1928. Austrian and German art historians, painters, journalists and restorers debated 'restoration concepts normally attributed to later decades.'[32]

An author largely ignored by all but historians is Manfred Holyoake, who in 1870 prefigured most of current conservation debate in a book aptly named *The Conservation of Pictures*. He defines conservation as 'the wider art of preserving as well as restoring the works.'[33] He recommends minimum intervention: 'The broad rule as to what is best to be done after the cleaning of a fine picture, is to make as few repairs as possible.'[34] He defines limits to cleaning: 'That which is foreign to the surface of a picture should be removed, so far as it may safely be done.'[35] He defines limits to reintegration: 'The moment restoration is suffered to proceed on fancy, the door is opened to discontent and disputations that never end.'[36] He is way ahead of current trends in accreditation of restorers, calling for 'a diploma as a guarantee to the public of his competency and knowledge on the subject,' and of the need for public involvement, 'the public like to know what is done to their pictures, and who does that which is done.'[37] He has ideas about the cost-benefit relation in heritage conservation: 'Not only does expenditure in putting a painting into condition produce its equivalent in the permanency it imparts to the duration of the pictures, but that very duration yields an interest like that of capital.'[38]

Who in the twenty-first century could disagree with his observation, 'The conserver has not always an intelligent employer?'[39]

Holyoake was a product of his time, rehearsing debates of his time, debates that have continued to this day. This shows that, even if progress in the development of ideas is real, it is very slow. What we often have presented to us is a succession of points in time that suggest an evolution. What we actually have is a number of points in time where someone whose name happens to get remembered said something, bits of which will be remembered. As we will see in the last section, the illusion of progress may merely be the result of people politely waiting in line to put their point of view.

Convergence

Conservators have always had something other than ethics on their mind. They have always worried about their status. When I joined the museum, object conservators were paid considerably less than painting conservators. As someone who worked in an objects studio, I could not see the justice in this. When I became a senior manager I tried, with other heads of conservation departments, to rectify this anomaly. The only way to do this was to use government grading systems that automatically valued management load above subject expertise and manual skill. This, among other factors, led to conservators striving to gain status and money by doing things other than treating objects with great skill, or helping objects to reveal their hidden stories.

Conservators, in a bid for managerial status, overwhelmed themselves with self-imposed administrative tasks such as condition reports. It became possible to gain respect and promotion by developing systems rather than healing objects. Under government pressure, senior managers used crude performance indicators such as number of objects 'treated,' treasuring throughput rather than difference made. As pressure of work built up, and the project culture kicked in, and the number of non-object-related tasks increased, the conservators slowly became less skilled. The highest paid became the least productive. As Lloyd-George had said of the British in 1940, 'Our people have become more sophisticated but less wise; intellectually more elaborately taught, but practically less competent.'[40] Obviously this is a compressed and exaggerated history, but it is one that could provide one explanation for the move away from complex and challenging treatments toward 'minimal intervention' that does not require recourse to a universal moral principle.

The other path towards higher status is the move towards recognized professional standing. The preamble to the 1981 UKIC Guidance for Conservation Practice declared the description of a professional standard as the first stage in satisfying official demands for a register of approved practitioners. Major advisory and funding bodies had called for a single conservation voice. The guidance was seen as the first step toward helping conservators 'speak and act with weight and authority' with 'a uniform attitude to the ethical practice of conservation.'[41] The notion of a profession is necessarily exclusive; there has to be a way of excluding people. If the profession is defined in terms of adherence to an ethical code, the obvious route to exclusion is by defining some behaviour as good and some behaviour as bad, then taking the moral high ground. The drive toward professionalism during the early 1980s probably introduced the moralistic judgement of treatment options that I did not notice in the early 1970s.

I have made the point that different branches of conservation may have different ethical bases. The only way to make those different thought systems appear united is with vague generalized statements of ethical behaviour. I have frequently remarked that conservators are convinced that there is one single conservation profession. Individually they may not understand nor care for the work and ethics of other groups or individuals that call themselves conservators, but they will always include them in their club as people with a common purpose. Titika Malkogeorgou, researching the ethnography of the V&A conservation department, has noted that, on the subject of ethics, conservators frequently contradict themselves, but never contradict one another.[42] This observation supports the view that conservators find it hard to use the bland guidance provided, but nonetheless see it as the basis for the desired goal of one united profession.

The motivation for union is the supposed status and respect brought about by a single powerful voice. As the dreams outlined in the 1980s have slowly come true in the UK with the formation of one major conservation institute, the trend has been for the differences to be ignored rather than overcome. New subjects that do not differentiate between object types or do not call for specialist manual skills have gained supremacy: preventive conservation, collections management, integrated pest management, disaster reaction, risk assessment. Concerns about climate change and sustainability will soon further satisfy the need.

Frank Hassard has noted that one of the main thrusts of the new professionalism, Continuing Professional Development (CPD), is detrimental to the sustainability of traditional practical skills.[43] CPD supposes a continuous change in the knowledge and skills needed for conservation practice, whereas traditional craft skills are by definition unchanging and unlikely to be picked up on a top-up course. As formal conservation training becomes shorter in time, but more demanding in curriculum content, the practical skills taught have to be simple and direct. So in instances where restoration is still allowable it will involve materials and skills unlike those used in the original manufacture. Indeed this approach will be defended as the only ethical option.

The move away from active restoration and the abandonment of traditional manual skills comes at an interesting time in the global development of heritage conservation. At the start of this discussion I cited four charters as influential. The three most recent charters place increasing importance on the intangible, and on relativistic interpretations of authenticity. In the majority of European and North American museums, collections care and conservation treatment have been based on unswerving respect for the original object. The originality of the object has been interpreted in terms of the very material present at the object's creation and the unchanged microscopic and macroscopic structure of that material.

This is not a universally held interpretation of authenticity. It probably has its origins in the reductionist scientific movement. In less materialistic cultures, sustainability of the intangible values and sustainability of relevant craft skills have been more important in defining authenticity. This stance allows regular physical maintenance, which leads to continuity of both appearance and utility. Earlier in this chapter I proposed this view of conservation as a hard-wired part of human nature, and the memes of the scientific paradigm as leading to aberrant behaviour. Not every culture has progressed so far from its basic nature. The growing acceptance of the intangible values associated with tangible heritage is not a sign of universal progress. It is a sign that others voices are at last being heard.

Since the early 1990s, conservators have become more sensitive to the value of the sacred. The need to consider spiritual, non-scientific, views on the care and treatment of objects are now built into the codes of practice. The notion that these views should predominate in some circumstances has spread and gained wide acceptance, encouraged by the work of Miriam Clavir.[44] The idea of conservation as 'the management of change' is gaining ground.[45] Where there is potential for rapid change it is accepted that maintenance is essential. Maintenance using like-with-like restoration, while it has never been completely outlawed, has been through a period of disfavour.[46] But where it has continued to be accepted, it carries with it the obligation to maintain practical skills. It is possible that the decay of objects in museums and historic houses is so slow that the decline in practical skills available for their conservation will not be a problem.

Summary

Some ideas about right and wrong in the treatment of cultural objects are innate. Ideas of what is possible come from the practical opportunities and constraints of materials in the form in which they are met by the conservator. Feelings of what is desirable are balanced by fears of public disapproval. What a majority think should be approved or disapproved will be determined by the memes that have spread most effectively at that time. Fashions in conservation spread slowly and unevenly, but the constraints of politics and economics favour the development of some modes of behaviour more than others. The drive to define a profession in terms of ethical behaviour, rather than knowledge and competence, has led to the belief that there is one set of ethics for all occasions. The tolerance of diversity is one way out of this situation.

Notes

1. See, for example, Jonathan Ashley-Smith, "Reversibility – Politics and Economics," *Reversibility: Does It Exist?, British Museum Occasional Paper 135*, eds. Andrew Oddy and Sara Carroll (1999) 129; Caroline Villers, "Post minimal intervention," *The Conservator,* Volume 28 (2004): 3.

2. Cesare Brandi, *Theory of Restoration* (Florence: Nardini Editore, 2005) (first published in Italian in 1963). Salvador Muñoz Viñas, *Contemporary Theory of Conservation* (Oxford: Elsevier Butterworth-Heinemann, 2005).
3. The Venice Charter: International Charter for the Conservation and Restoration of Monuments and Sites (1964) (http://www.international.icomos.org/e_venice.htm). The Burra Charter: The Australia ICOMOS Charter "For the Conservation of Places of Cultural Significance" (1999) (http://www.icomos.org/australia/burra.html). The Nara Document on Authenticity (1994) (http://www.international.icomos.org/naradoc_eng.htm) Yamato Declaration on Integrated Approaches for Safeguarding Tangible and Intangible Cultural Heritage. UNESCO (2004). (http://portal.unesco.org/culture/admin/file_download.php/Yamato_Declaration.pdf?URL_ID = 23863& filename = 10988742599Yamato_Declaration.pdf&filetype = application%2Fpdf&f ilesize = 35645&name = Yamato_Declaration.pdf&location = user-S/)
4. Michael Corfield, "Towards a Conservation Profession," *Conservation Today*, Preprints for the UKIC 30th Anniversary Conference (London: United Kingdom Institute for Conservation, 1988): 4.
5. The Murray Pease Report. "Report of the Murray Pease Committee: IIC American Group Standards of Practice and Professional Relations for Conservators," *Studies in Conservation*, Volume 9, No. 3 (August 1964): 116–121. American Institute for Conservation "AIC Code of Ethics and Guidelines for Practice" (Revised August 1994) (http://aic.stanford.edu/about/coredocs/coe/index.html). Canadian Association of Professional Conservators "Code of Ethics and Guidelines for Practice" (http://www.cac-accr.ca/english/e-CAC-about-Code-of-Ethics.asp). European Confederation of Conservator-Restorers' Organisations "E.C.C.O. Professional Guidelines II – Code of Ethics" (2003) (http://www.ecco-eu.info/index.php?container_id = 163&doc_id = 170).
6. "The Ten Commandments" feature in a story in Judaeo–Christian lore. *The Holy Bible Exodus*, Volume 20: 2–17.
7. Emily Post, *Etiquette (in society, in business, in politics and at home)* (New York: Funk and Wagnalls Company, 1922) available in replica edition (London: Cassell, 1969).
8. James Beck and Michael Daley, *Art Restoration: The Culture, the Business and the Scandal* (New York: W.W. Norton & Co., 1993) (http://www.artwatchinternational.org/need/bill.asp).
9. United Kingdom Institute for Conservation "Code of Ethics and Rules of Practice" (http://www.instituteofconservation.org.uk/groups/ukic/docs.html).
10. Edward O. Wilson, *Consilience: The Unity of Knowledge* (London: Abacus, 1999) 137.
11. George P. Murdock, "The common denominator of culture," *The Science of Man in the World Crisis*, Linton Ralph (New York: Columbia University Press, 1945).
12. Donald E. Brown, *Human Universals* (Philadelphia: Temple University Press, 1991). Wilson 162.
13. Leda Cosmides and John Tooby, "Cognitive adaptations for social exchange," *The Adapted Mind* (Oxford: Oxford University Press, 1992).
14. Susan Blackmore, *The Meme Machine* (Oxford: Oxford University Press, 1999) 6.
15. Richard Dawkins, *The Selfish Gene* (Oxford: Oxford University Press, 1976) (revised 1989), p. 192.

16. Martin E.P. Seligman, *Biological Boundaries of Learning* (New York: Appleton-Century-Crofts, 1972) 1–6.
17. Miriam Clavir, "The social and historic construction of professional values in conservation," *Studies in Conservation*, Volume 43 (1998): 1–8. Jonathan Ashley-Smith, "Restoration is an eleven letter word," *V&A Conservation Newsletter*, Volume 8 (June 1980). (http://www.jonsmith.demon.co.uk/AS_Family_Site/JAS_Site/publications.htm).
18. Jeremy Bentham, *A Table of the Springs of Action*. Digitized from Volume I of the 1843 Bowring edition of Bentham's works. (http://www.la.utexas.edu/research/poltheory/bentham/springs/springs.toc.html).
19. Richard Dawkins, *The Blind Watchmaker* (Harlow Essex: Longman, 1986).
20. R.B. Rutherford, *The Meditations of Marcus Aurelius: A Study* (Oxford: Clarendon Press, 1991).
21. Buddhism (http://www.sacred-texts.com/bud/index.htm); Islam (http://www.quranbrowser.com/); Christianity (http://www.biblegateway.com/).
22. Jonathan Ashley-Smith, "The Ethics of Conservation," *The Conservator*, Volume 6 (1982): 1.
23. Jonathan Ashley-Smith, "A Consistent Approach to a Varied Collection," *Restoration: Is it Acceptable? British Museum Occasional Paper 99*, ed. Andrew Oddy (London: British Museum Press, 1994), 89.
24. Canadian Association of Professional Conservators "Code of Ethics and Guidelines for Practice" (http://www.cac-accr.ca/english/e-CAC-about-Code-of-Ethics.asp).
25. Joyce Plesters and H. Ruhemann, *The Cleaning of Paintings* (London: Faber and Faber, 1968) p. 427.
26. Muñoz Viñas 212.
27. H.J. Plenderleith and Alfred Emil A. Werner, *The Conservation of Antiquities and Works of Art: Treatment, Repair and Restoration* (Oxford: Oxford University Press, 1971).
28. Although the Murray Pease Report was published by IIC in 1964, it was not considered an IIC publication. Moreover, it avoids the use of the word 'ethics.'
29. Hanna Jedrzejewska, *Ethics in Conservation* (Stockholm: Kungl Konsthogskolan, 1976), 5.
30. Andrew W. Oddy and Sara Carroll (eds.), *Restoration: Is it Acceptable? British Museum Occasional Paper 99* (London: British Museum Press, 1994). Andrew W. Oddy, ed., *Reversibility: Does It Exist? British Museum Occasional Paper 135* (London: British Museum Press, 1999).
31. Barbara Appelbaum, "Criteria For Treatment: Reversibility," *JAIC*, Volume 26, No. 2 (1987).
32. Michael von der Goltz, "Is it useful to restore paintings? Aspects of a 1928 discussion on restoration in Germany and Austria," *ICOM-CC Preprints 12th Triennial Meeting, Lyon, 29 August – 3 September 1999* (London: James and James, 1999), 200.
33. Manfred Holyoake, *The Conservation of Pictures* (London: Dalton & Lucy, 1870), 18.
34. Holyoake, p. 3.
35. Holyoake, p. 28.
36. Holyoake, p. 4.
37. Holyoake, pp. 73, 71.

38. Holyoake, p. 66.
39. Holyoake, p. 3.
40. D. Lloyd-George, "foreword in Vicomte de Mauduit," *They Can't Ration These* (1940) (republished London: Persephone Books Ltd, 2004), 11.
41. United Kingdom Institute for Conservation "UKIC Guidance for Conservation Practice" (1981).
42. Titika Malkogeorgou, personal communication, 17 August 2005.
43. Frank Hassard, "Continuing professional development and the surrender of culture to technology in the field of heritage preservation," *Preprints 15th Triennial Conference, International Council of Museums Conservation Committee (ICOM-CC), New Delhi, India, September 22–26 2008* (New Delhi: ICOM, 2008).
44. Miriam Clavir, *Preserving What is Valued: Museums, Conservation, and First Nations* (Vancouver: University of British Columbia Press, 2002).
45. See, for example, UNESCO World Heritage Committee, (http://www.newswire.com/ens/jul2006/2006-07-11-01.asp).
46. You can see signs of reluctant tolerance in the San Antonio Declaration. (http://www.icomos.org/docs/san_antonio.html).

3

The Aims of Conservation

Chris Caple

Introduction

All societies have objects they retain and cherish and in Europe, in the twenty-first century, that typically means placing them in a museum and letting conservators and other museum staff 'take care' of them. But we conservators are invariably focused on *how* and not *why* we are doing this. We spend our time talking to other conservators about 'ethical approaches' and obsess about the disparity between the different areas of conservation. We stand uncertain and mute as decisions are made in museums, universities and wider society that threaten the existence of the objects we care for and the institutions in which they reside. Do we have an accurate all-embracing view of conservation, a clear sense of purpose, a lucid series of aims, and can we articulate them in less than 500 pages? (e.g. Stanley-Price *et al.*, 1996).[1] If we cannot clearly and simply tell/convince society why we do what we do, what right do we have to intervene with society's most valued and treasured objects? In the following paragraphs I outline a basic series of aims for conservation. Do I accurately describe what conservation is and are these aims sustainable for the foreseeable future?

Maintain and enhance

Societies retain objects because they have value for the members of that society. These include religious values, aesthetic values, roles in ritual or ceremony, association with individuals venerated by that society and the value to educate or inform. Societies that retain objects invariably seek to 'maintain and enhance' the value of the object to that society. This may take many forms, such as participation in ceremonies, cleaning, repairing, use and anointing. Examples from cultures past and present include:

- **Objects reassembled**. Roman Samian vessels held together with lead rivets and seventeenth-century wineglasses held together with strips of lead.[2] No longer functional in the original sense, the objects are reassembled for heirloom value.

- **Objects repainted**. Aboriginal cave paintings,[3] the religious statues in French Catholic churches[4] and Māori buildings and objects[5] are repainted as a mark of respect to the spirits, saints or ancestors. The act of repainting rejuvenates the power of these objects.
- **Objects restored**. Objects, such as those made of Japanese *urushi*, are restored to their original appearance as a damaged or imperfect object would be considered to be disrespectful. Consequently, no differentiation is made between new and original material. Traditional rituals, tools and materials are used since the act of repair must also be performed in a respectful manner.[6]
- **Object storage and fumigation**. Some Native American peoples require certain objects to be handled by specific individuals and insect attack can only be treated with natural plant extracts that are not harmful to human beings.[7] These actions maintain the spiritual purity and power of the object, which is considered a 'living' being.

By the nineteenth century European and American societies had begun to appreciate that objects were more than commodities or symbols; they contained important information about the past, as articulated in the 1877 manifesto of the Society for the Protection of Ancient Buildings.[8] Consequently, 'maintenance and enhancement' for this society became the recovery, protection, cleaning, re-assembly and housing of objects in museums, art galleries, libraries and archives. Conservation was one of the terms that began to be applied to these activities as exemplified by the RIBA's 1865 booklet entitled *Conservation of Ancient Monuments and Remains*. However, the exact meaning of the term conservation has varied with each user and is related to their geographic and cultural origins and the type of artefacts that they 'conserve.'[9]

The 'aims of conservation' can be understood to refer to the 'purpose' or 'intentions' of conservation, what those who enact conservation seek to achieve. Evidence of purpose or intent comes in two forms; what is said (written) and what is done (conservation work carried out on objects or structures).

Modern society and museums

The seventeenth and eighteenth centuries (Age of Enlightenment) saw the giving of lectures (The Royal Society was established in 1660) and the publication of books that advanced ideas in a logical manner based on observation and classification of the physical (natural) world, such as *Systema Naturae* by Linnaes (published in 1735). From this point European and American societies have increasingly seen the world in scientific and logical terms; objects provide evidence (physical proof) about past and present-day societies; specimens exemplify the extent and nature of the natural world; devices demonstrate scientific principles, and works of art articulate emotions, ideas, aesthetics and explore symbolism and meaning in society. Preserved through collection, storage and display, objects, specimens, devices or works of art can be re-examined to reveal more information, and through public display they can potentially inform all members of society. These objects, specimens, devices and works of art constitute our proof, the physical evidence, for almost every facet of the development of humankind and almost every aspect of the forces of nature.

Archaeologists and anthropologists recognise objects as simultaneously existing in three forms: as functional artefacts (created to perform specific tasks), as symbols (culturally contexted meaning) and historic documents (record of the object's own past, its manufacture, use and existence as a functional artefact and symbol).[10,11] The objects collected by society into museums and archives are normally utilized as historic documents, to provide information (proof) about the past or other cultures.

Differing traditions within conservation

- Architects primarily focus on buildings; their aim is to maintain and enhance (preserve and restore) them, primarily as whole structures. They are aware that the costs of maintaining buildings (weatherproof and watertight) are high; thus, for a building to be maintained it must be used. Consequently, minor alterations damaging the building fabric to enable the provision of modern services (electrical and communication cables, water and sewerage pipes) are often perceived as necessary to ensure that the aim of preserving the building is achieved.
- For works of art, from oil paintings to sculpture, the primary focus is on the image providing visual stimulation, communicating or creating an emotion or feeling in the viewer. Conservators working on works of art primarily aim to maintain and enhance (restore) the original nature and quality of the image. Controversies in cleaning art, whether yellowing varnish on oil paintings or the grime on wall paintings, such as the Sistine Chapel, focus on the authenticity of the image and the aesthetic response of the viewers to it.[12]
- Archaeological and ethnographic conservators focus on maintaining (preserving) the existing artefact and enhance it through cleaning away dirt and corrosion to reveal further information about the object and its past. However, what constitutes evidence of the past has changed. The impressions of the organic materials, preserved in the minerals formed from the corrosion of an iron Viking sword from Sanday, provide evidence of the structure and composition of a tenth-century scabbard,[13] information that would have been cleaned away a generation before. This expanded understanding of what constitutes evidence of the past and the need to identify, record and preserve it has also become an increasing focus of the conservation of textiles.[14]

Thus what and how we seek to 'maintain and enhance' varies between objects – especially those in different traditions of conservation. It also changes with time. So although rooted in the wider requirements that society has for retaining objects, the 'aims of conservation' must be of a conceptual nature in order to allow for a number of differing conservation traditions and to avoid being made redundant by developing technology and increasing knowledge.

RIP triangle

Building on the 1984 ICOM-CC definition of a conservator-restorer, 'the activity of the conservator-restorer (conservation) consists of examination, preservation and conservation-restoration of cultural property,' I have previously suggested that conservation can be considered to have three competing aims, which together seek to maintain and enhance objects as historic documents (Figure 3.1):[15]

- **Revelation**. Cleaning and exposing the object, to reveal 'its original,' 'an earlier' or 'more meaningful' appearance. This appearance can be restored to give the observer, typically a museum visitor, a clearer visual impression of the object.
- **Investigation**. Researching, investigating and analysing the object to recover information about it. This may include visual observation, typological analysis, X-radiography, elemental or molecular analysis, even destructive analysis such as removing a metallographic section.
- **Preservation**. Maintaining the object in its present physical and chemical form, preventing any further deterioration, utilizing the stabilization processes of remedial (interventive) conservation and/or preventive conservation practices.

The balance of these aims forms a triangle, which defines the area in which activities can be described as conservation, and within which professional conservators work (Figure 3.1). Cleaning an object may aid its preservation, reveal the form of the object and uncover information about it.

If RIP accurately describes the aims of conservation, then even if you are simply repackaging objects in a store as a preservative action, the conscious act of ensuring the correct labelling of objects and boxes relates the object to its museum record (the accumulated information about the object) and enables it to be recovered for display, and is thus an act of conservation. Consciously performing such balanced actions means you do not need to be a qualified conservator to engage with the aims of conservation. Such an approach does not require substantial resources, only a clear understanding of why we, as a society, 'maintain and enhance' objects.

The Aims of Conservation

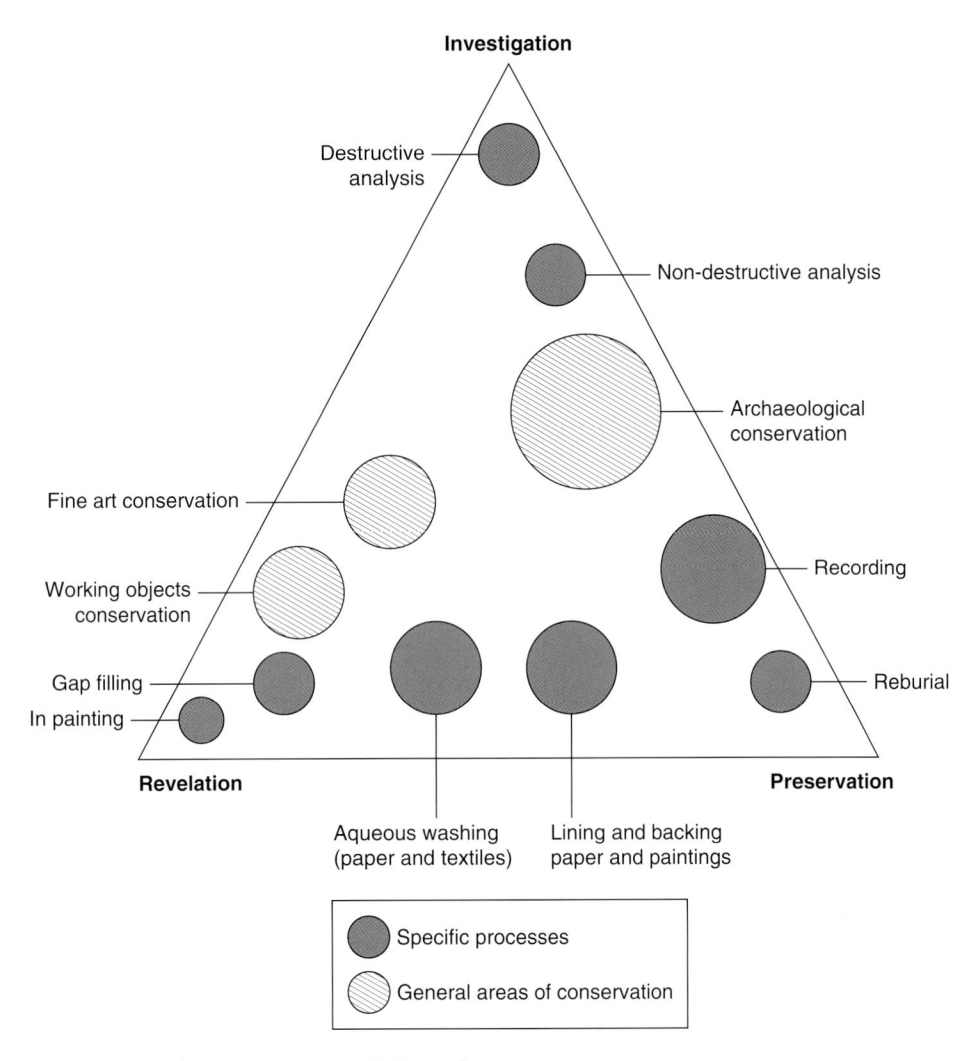

Figure 3.1 *The Conservation RIP Triangle.*

And yet...

The future holds a number of developments that cannot easily be reconciled with the aims of conservation outlined above.

- **Repatriation**. Museums, responding to the political and social pressure of 'native peoples,' are returning bones and 'religious' items from their collections, often for reburial. These objects and the unique information they contain are not being preserved and are permanently lost to the world of factual knowledge and scientific understanding.

- **Continued Collecting**. Logic tells us that we cannot afford to store (preserve) an ever-increasing number of objects. Will objects be sold off or will standards of care be lowered and which of these options should conservators advocate as they seek to meet the aims of conservation? Cheaper options such as written and pictorial records (virtual collections) fail to preserve the physical, re-examinable proof of the development of human kind and the forces of nature. Such records can be erroneous, faked and are limited to what we see and understand now. They are inherently unable to record what we might like to know in the future; how could we have proved Piltdown Man to be a fake if it had only been recorded as a picture?
- **Scientific Developments**. We are able to recover increasing amounts of information from objects, such as organic residues from ceramics or DNA from natural history specimens and archaeological bone. The best storage conditions for objects in order to preserve this information, such as under liquid nitrogen, are expensive and not compatible with display or access for other types of research.
- **Changing Function of Museums**. National Museums, Local Authorities and other organizations that own museum collections are increasingly concerned with short-term social needs to educate and entertain – measured in terms of museum visitor numbers. The *raison d'être* of collections as proof/information about the past is forgotten and resources increasingly moved away from preservation and research (investigation) to display (revelation).

As the ideals of the Age of Enlightenment are lost and social values are increasingly focused on mass entertainment, increasing personal wealth and fundamental religious principles, society will redefine why it keeps the objects of the past. Will conservation need to redefine its aims, or if society wishes to 'maintain and enhance' its objects in a way that no longer reveals, investigates and preserves them, does what we do cease to be conservation?

Bibliography

1. Nicholas Stanley-Price, M. Kirby Talley Jr. and Alessandra Melucco Vaccaro, *Historical and Philosophical Issues in the Conservation of Cultural Heritage* (Los Angeles: The Getty Conservation Institute, 1996).
2. Hugh Willmott, "A group of 17th century glass goblets with restored Stems: Considering the archaeology of repair," *Post Medieval Archaeology*, Volume 35 (2001): 96–105.
3. L. Maynard, "Restoration of Aboriginal Rock Art – the moral problem," *Proceedings of the National Seminar on the Conservation of Cultural Materials* (Perth: AICCM, 1973).
4. Tiamat Molina and Marie Pincemin, "Restoration Acceptable to Whom?" *Restoration: Is It Acceptable?* (British Museum Occasional Paper No. 99) (London: British Museum, 1994).
5. Gerry Barton and Sabine Weik, "Māori carvings: Ethical considerations in their conservation," *ICOM-CC 7th Triennial Meeting, Copenhagen 1984* (Copenhagen: ICOM-CC, 1984).

6. Shin Yakihashi, "The preservation and handing down of traditional urushi art techniques in Japan," *Urushi* (Marina del Rey: Getty Conservation Institute, 1988).
7. Ann Drumheller and Marian Kaminitz, "Traditional Care and Conservation, the merging of two disciplines at the National Museum of the American Indian," *Preventive Conservation Practice, Theory and Research: 1994 IIC Ottawa Congress* (London: IIC, 1994).
8. William Morris, "Manifesto of the Society for the Protection of Ancient Buildings," *Historical and Philosophical Issues in the Conservation of Cultural Heritage* (Los Angeles: The Getty Conservation Institute, 1996) (first published in 1877).
9. Elizabeth Pye, *Caring for the Past* (London: James and James, 2001), pp. 26–30.
10. Susan M. Pearce, *Interpreting Objects and Collections* (London: Routledge, 1994), p. 12.
11. Chris Caple, *Objects: Reluctant Witnesses to the Past* (London: Routledge, 2006) 6–11.
12. Gerry Hedley, "Cleaning and meaning: The Ravished image reviewed," *The Conservator*, Volume 10 (1986): 2–6.
13. Olwyn Owen and Magnar Dalland, *Scar, A Viking Boat Burial on Sanday, Orkney* (Edinburgh: Historic Scotland, 1999) 103–112.
14. Mary Brooks, Alison Lister, Dinah Eastop, and Tarja Bennett, "Artefact or Information? Articulating the Conflicts in Conserving Archaeological Textiles," *Archaeological Conservation and its Consequences: IIC 1996 Copenhagen Congress* (London: IIC, 1996).
15. Chris Caple, *Conservation Skills: Judgement, Method and Decision Making* (London: Routledge, 2000) 31–35.

4

The Reconstruction of Ruins: Principles and Practice

Nicholas Stanley-Price

Introduction

Reconstruction has always been one of the most controversial issues for those with an interest in the material evidence of the past. The urge to make whole again a valued building or work of art that is incomplete is a very strong one, similar in some ways to the urge to improve or correct someone else's text. Both involve a strong desire to see an object that is complete and integral to one's own satisfaction, rather than tolerate a creative work that has been diminished in its intelligibility.

The idea that the object may have a greater value in its incomplete state than if it is reconstructed, runs counter to this strong compulsion. Yet that idea has been central to much of the theory of conservation and restoration that developed primarily in the Western world and has subsequently been diffused worldwide.[1] The core of Western conservation theory is epitomized in the question as to how far restoration should be taken.

Different attitudes towards this fundamental question have given rise to some of the most notorious controversies in conservation. For instance, disagreements over the extent to which paintings at the National Gallery of London should be cleaned, and what methods should be used, led to official Commissions of Enquiry in 1850 and 1853 and remarkably, a century later, were revived following the criticisms by Cesare Brandi and others of what they considered the Gallery's excessive cleaning of early paintings.[2] Another example is John Ruskin's critique in the nineteenth century of the 'stylistic restoration' of historic buildings that aimed at reviving earlier styles rather than respecting the age-value and patina that a building had accumulated through time.[3]

A number of important concepts, such as reversibility (or, better, re-treatability) and minimum intervention, are at the heart of an ever-growing library of Codes of Ethics and Charters. Nevertheless, there are no textbook rules about when restoration should be carried out or how far it should go. Instead, each case is deemed to be different and must be judged on its merits.[4] This is perhaps what gives conservation/restoration much of its perpetual fascination.

In order to examine the question here, I consider the reconstruction of ruins, which represents in many respects an extreme example of restoration. In order to define the question as clearly as possible, I limit the discussion to buildings from the past whose existence was known primarily from their excavated remains before being reconstructed. In other words, although there may be other references – literary, folkloric or pictorial – to their previous existence, it is mainly through their insubstantial visible remains that they have become known again.

I have deliberately limited the argument in this way, in the hope of avoiding the confusion that could be introduced by including other types of building reconstruction. I do not consider here buildings that have been reconstructed immediately following a natural disaster or a war. These differ because there usually exists ample documentary evidence of the destroyed buildings. Examples include the main hall of the Horyu-ji Temple at Nara in Japan, burnt in 1949; the Campanile in the Piazza di San Marco, Venice that suddenly collapsed in 1902; the Old Town of Warsaw; the Frauenkirche in Dresden destroyed during WWII; and the Old Bridge at Mostar destroyed during the recent war in the Balkans.

Nor do I consider projects to reconstruct historic buildings that are known to have existed in the distant past but for which only sparse literary and pictorial references survive. (This practice is often referred to as re-creation, if the result is highly conjectural.) The strong trend, especially in former Communist states, towards reconstructing such vanished buildings, often on the basis of flimsy documentary evidence of their original appearance, is generating its own critiques.[5] Several of the arguments adduced below are relevant to these cases, but they are not the focus of this chapter.

So the question that is posed here is: When should such excavated and incomplete buildings be reconstructed to a state similar to how they might once have appeared? The chapter examines in turn the following questions: What widely accepted principles are there concerning reconstruction? How has the practice of reconstruction been justified (whatever the accepted principles may be)? What are the arguments against it? And finally, in the light of arguments for and against, what principles can be proposed to help guide issues of reconstruction?

Principles enshrined in conventions and charters

In international legislation and guidelines, the answer to the question as to whether incomplete buildings should be reconstructed is clear. It is strongly discouraged.

At the highest level of international consensus, the obligations of UNESCO's World Heritage Convention (1972) are legally binding on the states party to it; the number of states party is in fact the highest of any UNESCO Convention.

The Operational Guidelines for the Implementation of the World Heritage Convention address the question of reconstruction of buildings as follows:

> In relation to authenticity, the reconstruction of archaeological remains or historic buildings or districts is justifiable only in exceptional circumstances. Reconstruction is acceptable only on the basis of complete and detailed documentation and to no extent on conjecture.[6]

To repeat, the obligations of international conventions of the United Nations are legally binding on their states party. Charters, on the other hand, tend to have an exhortatory role in encouraging professionals to adopt commonly agreed principles in their work. The content and eventual impact of a Charter depends, de facto, on the authority of those who drafted and approved it, and thence its acceptability to the professional field in general. Several Charters in conservation have addressed the question of reconstruction of sites on the basis of their archaeological remains.

For example, the influential Charter of Venice (1964) states with regard to the reconstruction of archaeological sites (Article 15): 'all reconstruction work should however be ruled out. Only anastylosis, that is to say, the reassembling of existing but dismembered parts, can be permitted.'

The strong presumption against reconstruction expressed in the Operational Guidelines for the Implementation for the World Heritage Convention and in the Venice Charter is echoed in many subsequent documents. For instance, the revised version (1999) of the Burra Charter of Australia ICOMOS, originally developed for the Australian context but cited much more widely as a coherent set of guidelines, states:

> Article 1.8. Reconstruction means returning a place to a known earlier state and is distinguished from restoration by the introduction of new material into the fabric.
>
> Article 20. Reconstruction.
> 20.1. Reconstruction is appropriate only where a place is incomplete through damage or alteration, and only where there is sufficient evidence to reproduce an earlier state of the fabric. In rare cases, reconstruction may also be appropriate as part of a use or practice that retains the cultural significance of the place.
>
> 20.2. Reconstruction should be identifiable on close inspection or through additional interpretation.

The language of the Venice Charter is uncompromising in proposing what constitutes acceptable reconstruction on archaeological sites ('the reassembling of existing but dismembered parts'). But the interpretation of reconstruction in the

Burra Charter (Article 1.8 above) as being 'distinguished from restoration by the introduction of new material into the fabric' is at variance with the Venice Charter and with common usage outside Australia. There must be few restorations that do not require the introduction of any new material. If the Burra Charter definitions were to be widely adopted outside Australia for where they were developed, they could not fail to cause confusion. For instance, the current long-term project on the Acropolis of Athens would have to be characterized as a reconstruction, a term that would be rejected by the Greek authorities.[7]

What is common to all such documents, whether they are international conventions or charters produced by groups of professionals, is that reconstruction constitutes an exceptional case and should be carried out only when there exists sufficient primary evidence. As the World Heritage Operational Guidelines state, reconstruction is 'acceptable only on the basis of complete and detailed documentation and to no extent on conjecture.'

In reality, the strictures of these international documents have prevented neither the continued practice of reconstruction nor the inscription of sites with reconstructed buildings on the World Heritage List nor new reconstructions on sites already so inscribed. It is striking that a recent volume of essays on site reconstructions contains but one reference to the Charter of Venice, and mentions World Heritage only in the context of sites inscribed on the List that feature reconstructions, e.g. the prehistoric Aztec Ruins and Mesa Verde in the USA.[8] It is as if such reconstructions are justified for their public interpretation value whether or not they meet the criteria of international restoration documents.

In fact, and not only in the USA, despite the almost universal consensus of the charters against reconstruction unless firmly based on evidence, it still holds a strong appeal – both for cultural heritage managers and for the public. So how has the reconstruction of buildings known from their excavated remains been justified, and what are the arguments against the practice?

Justifications for reconstruction

A number of justifications have been given for the reconstruction of buildings that are known primarily from excavated evidence.[9] These include:

1. **National symbolic value**. The building played an important role in the country's history, or was associated with an outstanding figure.

   I give only two examples of what is probably the commonest impulse towards reconstruction, both of them from former capitals in their countries. Because of its important role in what was the capital of Virginia until 1775, the Governor's Palace (1706–1791) was the first major building to be reconstructed after the project to 'restore' Colonial Williamsburg began in 1927.[10] Much of today's reconstructed Palace interior is quite hypothetical, but the footprint for the reconstructed building was established by non-expert excavation in the 1920–1930s to expose the original foundations (the first professional archaeologist was not appointed at Williamsburg until as late as 1957).

In Japan, at the eighth-century AD Heijô Palace site of Nara, a place of immense symbolic value in Japanese history, the insubstantial traces of the wooden buildi\ngs revealed by excavation have led to full-scale reconstructions of the Suzakmon Gate (1990–1997) and, since 2001, of the Daigokuden Hall of the Palace.

2. **Continuing function or re-use**. The reconstructed building can continue to serve its previous function or makes possible a new, different function.

   Rarely are excavated buildings reconstructed to serve their previous or original function. The principal exceptions are Greek and Roman theatres and other places of performance. Buildings that have been extensively reconstructed from archaeological evidence to serve new functions would include the Stoa of Attalus in the Athenian Agora, reconstructed in 1953–1956 to serve as a museum, store and workspace for finds from the continuing excavations there.[11]

3. **Education and research**. The process of reconstruction can be a rewarding research project, and the resulting building is an important didactic tool for visitors. 'Visitors love them.'

   If interpreted broadly, this justification holds true for the great majority of reconstructed sites. Whatever the primary motivation for it, a reconstructed building has the potential to have a high educational and research value. The very process of researching, testing and building unfailingly leads to a better understanding of the past by specialists. Non-specialists benefit from the new knowledge accumulated during the process and from viewing the built embodiment of it. The many reconstructions of timber buildings based upon archaeological evidence in the USA, northwest Europe and Japan exemplify the combined research and popular education roles of reconstructions.

4. **Tourism promotion**. A reconstructed building can attract tourism and thus generate income for the public or private authorities that manage it.

   The massive reconstruction of pre-Hispanic sites in Mexico, Guatemala, Belize and Bolivia (Tiwanaku) in the 1950s and 1960s aimed to promote tourism while also demonstrating national pride in the pre-Colombian past.[12] The motivation behind the proposed reconstruction of the Hwangnyongsa Temple in Gyeongju (Republic of Korea) is first and foremost the economic development of the city, especially through increased tourism, and not its potential re-use as a Buddhist temple.[13]

5. **Site preservation**. Reconstruction, by showing that the site is being actively used, helps protect it from development pressures; alternatively, it may serve to stabilize precarious ruined structures.

If a salvage excavation has taken place in advance of commercial development, reconstructing the building whose foundations have been excavated can prevent the alternative development going ahead.[14] A classic case of reconstruction (or reconstitution as he called it) being justified in order to stabilize excavated ruins is Arthur Evans' work at Knossos.[15] In fact, as C. Palyvou perceptively observes,[16] it was Evans' concern for preservation through reconstruction that led to his interest in site presentation (aided also by his communication qualities as a journalist), rather than the more common path of a concern for site presentation leading to reconstruction.

If these points summarize some of the main justifications that have been cited for reconstructing buildings from excavated remains, what are the arguments against this practice?

Arguments against reconstruction

A. *The evocative value of ruined buildings*. A ruined building left as it is can be more evocative of the past than that same building reconstructed.

The romantic appeal of ruins has been extensively written about,[17] if sometimes rather simplistically attributed to nostalgia for the past, which is supposedly characteristic of the European Romantic tradition. But the creative role of ruins in inspiring art, literature, and music cannot be discounted, nor the deliberate retention of ruins as memorials to tragic events. The preservation as a ruin of the A-Bomb Dome at Hiroshima is one example from outside Europe.

B. *The difficulty (impossibility?) of achieving authenticity*.[18] Reconstructed buildings are de facto new buildings, tending to reflect the culture and times of their creators, rather than being faithful reproductions of the original.

Very few reconstructions from excavated remains would meet the standard requirement of the Charters that they be based on full and complete documentation. It is hard to see how excavated remains alone could provide that. Because reconstructions do involve conjecture to a greater or less degree, the tendency will be for their architects to be unconsciously prone to other influences. Thus the influence of Beaux-Art ideals has been noted in the reconstructed Capitol building at Colonial Williamsburg and as a possible inspiration for Evans' use of colour in the Knossos reconstructions.[19,20] But the latter seem also to have been strongly influenced by contemporary Art Deco styles (Figure 4.1).[21]

Figure 4.1 *North Lustral Basin, Knossos, Greece as restored by Arthur Evans in 1929. Photo reproduced with permission from the Ashmolean Museum, Oxford.*

C. *The ethical issue of conveying erroneous information.* Inaccurate reconstructions can mislead the professional and lay publics unless identified as such.

   Despite the laudable justification of education and research goals (see point 3 above), if the reconstruction is inaccurate or simply wrong, both scholars and the lay public can be misled if not warned. The use of comparative evidence from other pre-Colombian sites for reconstructing Pyramid B at Tula in Mexico (Figure 4.2) led astray future scholars who were unaware of what had been reconstructed and how.[22] If professionals can be misled, what false impressions are non-specialist visitors to gain unless informed as to what has been reconstructed on a conjectural basis?

D. *The destruction of original evidence.* Many reconstructions have either destroyed or rendered inaccessible the evidence on which they are based, to the detriment of future scientific research.

   The reconstruction of buildings *in situ* on their original foundations, however credible it may be, is likely to limit the options for future research as ideas change. The ICOMOS *Charter for the Protection and Management of the Archaeological Heritage* (1990), Article 7, evidently has this risk in mind: 'Where possible and appropriate, reconstructions should not be built immediately on the archaeological remains and should be identifiable as such.' The horizontal displacement of any reconstruction work to another site as 'experimental archaeology' avoids this problem, as does 'vertical displacement' to some extent – I refer to the practice in Japan of leaving a layer of earth or concrete to separate the original subsurface remains from the foundations of the reconstruction.[23]

Figure 4.2 *Pyramid B, Tula, Mexico, as restored by Jorge Acosta, 1941.*

E. *The disruption of landscape values*. A reconstructed building in an otherwise ruined landscape distorts visual and spatial relationships.

If only one or two buildings are reconstructed on an otherwise 'flat' site, they tend to influence visitors' 'desire lines' (preferred circulation routes around the site). The reconstruction may enhance an appreciation of the original form of those particular buildings but the inequalities of scale will risk diminishing an understanding of the site as a whole. The monumental scale of the reconstructed Stoa of Attalus in the Athens Agora, already referred to (see point 2 above), the Gymnasium of the Baths at Sardis (Figure 4.3) and the Temple of Hatshepsut at Luxor exemplify this phenomenon.

Figure 4.3 *Gymnasium of Baths at Sardis, Turkey, as reconstructed in 1964–1973.*

F. *Distorted site interpretation.* The complexities of sites with a long history are obscured if they are reconstructed to feature a single period.

> In technical terms it is relatively easier to reconstruct to a single period, but the evidence of other periods may have to be sacrificed. At Knossos 'the casual visitor – and often even the specialist – can forget that Knossos is the largest Neolithic site on Crete…and…is one of the two largest Greek and Roman sites on the island.'[24] On the Acropolis of Athens, almost all evidence of post-Classical building had already been demolished in the nineteenth century as part of the post-Independence glorification of the remains of Classical Greece, thus facilitating the current project.[25] In other cases, political pressures may require a specific historical occupation phase to be emphasized on a multi-period site.[26]

G. *Cost.* Reconstruction projects tend to be very expensive and often can only be financed by the political authorities who insist they be undertaken.

> Without the support of a Rockefeller (who financed the plan to restore Colonial Williamsburg), it tends to be public authorities, using public funds, who make possible major reconstruction projects. So the decision to undertake them, and the criteria that define their scope and result, are usually not

those of professional heritage managers. Moreover, the subsequent maintenance costs are often not taken into account, and the costs of reconstructed sites tend to reduce the budgets available for other, less spectacular sites. An extreme case is the lavish reconstruction of Babylon, undertaken for political reasons while Iraq was engaged in a long-term and costly war with its neighbour Iran.[27] In a different kind of war, B. Mackintosh describes several battles, some successful and some not, fought by the National Park Service (NPS) in the USA to counter reconstruction projects advocated by Congressional representatives in their home districts.[28] The very popularity of the conjectural restorations of Colonial Williamsburg from their earliest results created amongst members of the public expectations that sites would be reconstructed, even where the evidential basis was lacking. Politicians did not hesitate to exploit their populist appeal and to make the necessary funds available, despite the official NPS policy or the views of the professionals.

Towards some principles for site reconstruction

On this controversial topic, it is difficult to propose guidelines – the gulf that exists between the statements of Charters and the World Heritage Convention guidelines and actual practice demonstrates this point. Nevertheless, in this concluding section I try to propose some principles. They take into account the previous discussions of justifications usually made for reconstruction and of arguments against it.

1. A reconstructed building – if based primarily on excavated evidence – must be considered a new building (reconstruction as a creative act).

2. Reconstruction of one or more buildings is to be considered only if the values (including the landscape value) of a site will be better appreciated than if the buildings are left in a ruined state (the ruin as a source of inspiration or as a memorial).

3. The surviving evidence for the former building must be fully documented in such a way that this record is always available in the future (a scientific and ethical obligation to record for posterity).

4. The surviving evidence for the former building, or for different historical phases of it, must not be destroyed or made inaccessible by the very act of reconstructing it (a scientific obligation to allow (built) hypotheses to be verified or rejected).

5. The evidence – its strengths and its limitations – for the reconstructed form must be interpreted clearly to all visitors (an ethical obligation not to mislead or misinform the public).

6. Buildings that have been wrongly reconstructed in the past could, on a case-by-case basis, be preserved as they are (reconstructions as part of the history of ideas).

It seems axiomatic that reconstructions of the kind described here are to be considered new buildings (as they are by contemporary architects who adopt bold solutions for adapting old buildings). They are not incomplete old buildings that have been 'restored to their former glory,' in the phrase beloved by the media. How many reconstructions have even attempted really to reproduce the conditions that are assumed to have obtained in the past? Criticisms of the 'too-clean Williamsburg' are well known and could be applied to all reconstructed sites. Evans' use of colour at Knossos is an exception to the general rule of non-painted architectural reconstructions in Classical lands. Significantly, Evans' colours were later toned down in the 1950s in accordance with changing taste, but have now been revived as part of the conservation project that considers Evans' work as part of the history of the site.[29] So, in short, reconstructions are new buildings; they do not reproduce original conditions.

The obligation to record and preserve evidence for future investigators must be inherent to any field of study that considers itself scientific. So any reconstruction should avoid impact on the original remains by means of either vertical or horizontal displacement (see D above). Equally, a reconstruction should aim at respecting the integrity of a building that has evolved through time. The removal of the remains of any one phase in the interests of the reconstruction of other phases must be justified and fully documented.

The requirement to convey to visitors accurate information about the fidelity of a reconstruction to the current state of knowledge seems paramount. Knowingly to convey inaccurate information without disclosure is unethical (or actually criminal) in other spheres of communicating with the public. Why should conjectural reconstructions be exempt from this requirement? The standard criterion in restoration of 'visibility of the intervention' applies here. It can be met either by employing subtle differences in the technique or texture of materials or more strikingly by using quite modern materials, perhaps reproducing only the volumes of the vanished buildings and not their solid form (i.e. volumetric reconstruction, as practiced for example at Benjamin Franklin's House in Philadelphia, the Forges St Maurice industrial installation in Québec, and the Temple of Apollo at Veii, on the northern outskirts of Rome).

A different argument can be made for retaining erroneous reconstructions carried out in the past, on the basis that they possess their own value in reflecting the history of taste and ideas (as in Evans' work at Knossos). A parallel exists with the restoration of antique sculpture, for which there is a value in retaining previous restorations even though erroneous.[30]

Conclusion

There is no doubt that the international normative documents and the ever-growing number of Charters guiding conservation practice have had a strong influence on conservation practice. But within the built heritage field the particular case of reconstruction exhibits a clear divergence between principles and practice.

In this chapter I have attempted to summarize some of the justifications that have been used for reconstructing buildings now known mainly from their excavated remains, and also some of the arguments against this practice. The hard line taken against reconstruction in the normative documents must stem from experience; in other words, a consensus has developed among professionals that the arguments against outweigh the justifications for. And yet vanished buildings continue to be reconstructed. Is there a way out of this paradox?

One way out lies in responding differently to the enormous popular appeal of reconstructed buildings. The advent of multimedia and virtual realities makes it possible to explore competing hypotheses about the past without requiring any intrusion into the original physical remains on-site. The high costs associated at present with the development of such projects will decline as technology evolves. Thus a visit to the 'real thing' in the field, appropriately conserved and interpreted as found, will be a test of the credibility of the electronically generated image of the past. An ability to appreciate the authenticity of the past depends in the end on the observer, and not on the observed. Or, put another way, it is the visitor who should be treated, and not the building.[31]

Notes

1. As it has spread, the philosophy of 'conserve as found' has come into conflict with traditions that provide for the regular renovation of buildings of continuing religious or other functions. It is now more widely admitted that it is the preservation of the spiritual values of such buildings ('living heritage') that is more important than conservation of their physical fabric alone.
2. See various readings in Part VI "Cleaning Controversies," *Issues in the Conservation of Paintings*, eds. D. Bomford and M. Leonard (eds) (Los Angeles: The Getty Conservation Institute, 2004) 425–547.
3. See introduction and readings in Part V "Restoration and anti-restoration," *Historical and Philosophical Issues in the Conservation of Cultural Heritage*, eds. N. Stanley Price, M.K. Talley, Jr. and A. Melucco Vaccaro (Los Angeles: The Getty Conservation Institute, 1996) 307–323.
4. See the different contributions in W.A. Oddy, ed. *Restoration: Is It Acceptable?*, British Museum Occasional Paper 99 (London: British Museum Press, 1994) and in *Faut-il Restaurer les Ruines?*, (Actes des Colloques de la Direction du Patrimoine.) Entretiens du Patrimoine (Paris: Picard, 1991).
5. For example, H. Stovel, "The Riga charter on authenticity and historical reconstruction in relationship to cultural heritage: (Riga, Latvia, October 2000)," *Conservation and Management of Archaeological Sites*, Volume 4. Number 4, (2001): 240–244; N. Dushkina, "Reconstruction and its Interpretation in Russia – 2," *Proceedings of the*

Scientific Symposium, Session II, paper 12, ICOMOS 15th General Assembly, Xi'an, China, 17–21 October 2005, accessed 6 February 2007, www.international.icomos.org/xian2005/papers.htm; and J. Pirkovic, "Reproducing lost monuments and the question of authenticity," *Varstvo spomenikov*, Volume 40 (2003): 209–221.

6. *Operational Guidelines for the Implementation of the World Heritage Convention* (Paris: UNESCO, revised 2005) §86. The wording is almost identical in the previous version of the Operational Guidelines concerning authenticity, with the significant addition of the words 'of the original': '(the Committee stressed that reconstruction is only acceptable if it is carried out on the basis of complete and detailed documentation of the original and to no extent on conjecture)' (Paris: UNESCO World Heritage Committee, 1998) §24(b) (I).

7. F. Mallouchou-Tufano, "Thirty years of anastelosis work on the Athenian Acropolis, 1975–2005," *Conservation and Management of Archaeological Sites*, Volume 8, Number 1, (2006): 27–38.

8. J.H. Jameson, ed., *The Reconstructed Past. Reconstruction in the Public Interpretation of Archaeology and History* (Walnut Creek: Altamira Press, 2004).

9. See, for example, J.M. Fitch, *Historic Preservation. Curatorial Management of the Built World* (Charlottesville: University Press of Virginia, 1990); P.G. Stone and P.G. Planel, eds., *The Constructed Past. Experimental Archaeology, Education and the Public*, One World Archaeology 36 (London: Routledge, 1999); and J.H. Jameson, ed., *The Reconstructed Past. Reconstruction in the Public Interpretation of Archaeology and History* (Walnut Creek: Altamira Press, 2004).

10. M.R. Brown III and E.A. Chappell, "Archeological Reconstruction and Authenticity at Colonial Williamsburg," *The Reconstructed Past. Reconstruction in the Public Interpretation of Archaeology and History*, ed. J.H. Jameson (Walnut Creek: Altamira Press, 2004) 47–63.

11. H.A. Thompson, *The Stoa of Attalos II in Athens*, Excavations of the Athenian Agora Picture Book no. 2 (Athens: American School of Classical Studies, 1959).

12. A. Molina-Montes, "Archaeological Buildings: Restoration or Misrepresentation," in ed. E.H. Boone, *Falsifications and Misreconstructions of pre-Columbian art, Dumbarton Oaks*, 14–15 October 1975, (Washington, DC: Dumbarton Oaks Institute of Meso-American Studies, 1982) 125–141; D. Schávelzon, *La Conservación del Patrimonio Cultural en América Latina. Restauración de Edificios Prehispánicos en Mesoamérica: 1750–1980* (Buenos Aires: Instituto de Arte Americano e Investigaciones Estéticas "Mario J. Buschiazzo," 1990).

13. H-S. Kim, "Utilization Plan of Hwangnyongsa Temple after Reconstruction," *Preprints of International Conference on Reconstruction of Hwangyongsa Temple, April, 28–April 29, 2006, Gyeongju-si, Korea* (Seoul: National Research Institute of Cultural Heritage, 2006) 385–401.

14. K. Okamura and R. Condon, "Reconstruction Sites and Education in Japan: a Case Study from the Kansai Region," *The Constructed Past. Experimental Archaeology, Education and the Public*, One World Archaeology 36, eds. P.G. Stone and P.G. Planel (London: Routledge, 1999) 63–75.

15. A.E. Evans, "Works of reconstitution in the palace of Knossos," *Antiquaries Journal* Volume 7 (1927): 258–267.
16. C. Palyvou, "Architecture and Archaeology: The Minoan Palaces in the Twenty-first Century," *Theory and Practice in Mediterranean Archaeology: Old World and New World Perspectives*, Cotsen Advanced Seminars 1, eds. J. K. Papadopoulos and R.M. Leventhal (eds) (Los Angeles: The Cotsen Institute of Archaeology, University of California at Los Angeles, 2003), 205–233.
17. For example, C. Woodward, *In Ruins* (Vintage, 2002).
18. H. Schmidt, "The impossibility of resurrecting the past: Reconstructions on archaeological excavation sites," *Conservation and Management of Archaeological Sites*, Volume 3, Number 1–2 (1999): 61–68.
19. C.R. Lounsbury, "Beaux-arts ideals and colonial reality: the Reconstruction of Williamsburg's Capitol 1928-1934," *Journal of the Society of Architectural Historians*, 49.4 (1990): 373–389.
20. C. Palyvou, "Architecture and Archacology: the Minoan Palaces in the Twenty-first Century," *Theory and Practice in Mediterranean Archaeology: Old World and New World Perspectives*, Cotsen Advanced Seminars 1, eds. J.K. Papadopoulos and R.M. Leventhal (Los Angeles: The Cotsen Institute of Archaeology, University of California at Los Angeles, 2003) 218–219.
21. See for instance the striking photograph of the North Lustral basin at Knossos as restored in 1929 reproduced here as Figure 4.1.
22. A. Molina-Montes, "Archaeological Buildings: Restoration or Misrepresentation," *Falsifications and Misreconstructions of pre-Columbian art, Dumbarton Oaks, 14–15 October 1975*, ed. E.H. Boone (Washington, DC: Dumbarton Oaks Institute of Meso-American Studies, 1982) 125–141.
23. H. Kanaseki, "Reconstructing a Ruin from Intangible Materials," *Nara Conference on Authenticity*, UNESCO World Heritage Centre, Agency for Cultural Affairs Japan, ICCROM, ICOMOS, ed. K.E. Larsen (Trondheim: Tapir, 1995) 337–338; K. Okamura and R. Condon, "Reconstruction Sites and Education in Japan: A Case Study from the Kansai Region," *The Constructed Past. Experimental Archaeology, Education and the Public*, One World Archaeology 36, eds. P.G. Stone and P.G. Planel (London: Routledge, 1999) 63–75.
24. J.K. Papadopoulos, "Knossos," *The Conservation of Archaeological Sites in the Mediterranean Region: an International Conference organized by the Getty Conservation Institute and the J. Paul Getty Museum, 6–12 May 1995*, ed. M. de la Torre (Los Angeles: Getty Conservation Institute, 1997) 115.
25. F. Mallouchou-Tufano, "Thirty years of anastelosis work on the Athenian Acropolis, 1975–2005," *Conservation and Management of Archaeological Sites*, Volume 8, Number 1 (2006): 27–38.
26. For example, A. Killebrew, "Reflections on a Reconstruction of the Ancient Qasrin Synagogue and Village," *The Reconstructed Past. Reconstruction in the Public Interpretation of Archaeology and History*, ed. J.H. Jameson (Walnut Creek: Altamira Press, 2004) 127–146.

27. R. Parapetti, "Recenti Interventi sul Patrimonio Archeologico in Iraq," *Restauro*, Volume 19, Number 110 (1990): 94–102.
28. B. Mackintosh, "National Park Service Reconstruction Policy and Practice," *The Reconstructed Past. Reconstruction in the Public Interpretation of Archaeology and History*, ed. J.H. Jameson (Walnut Creek: Altamira Press, 2004) 65–74.
29. C. Palyvou, "Architecture and Archaeology: the Minoan Palaces in the Twenty-first Century," *Theory and Practice in Mediterranean Archaeology: Old World and New World Perspectives*, Cotsen Advanced Seminars 1, eds. J.K. Papadopoulos and R.M. Leventhal (Los Angeles: The Cotsen Institute of Archaeology, University of California at Los Angeles, 2003), 227; J.K. Papadopoulos, "Knossos," *The Conservation of Archaeological Sites in the Mediterranean Region: an International Conference organized by the Getty Conservation Institute and the J. Paul Getty Museum, 6-12 May 1995*, ed. M. de la Torre (Los Angeles: Getty Conservation Institute, 1997) 116.
30. G. Vaughan, "Some Observations and Reflections on the Restoration of Antique Sculpture in the Eighteenth Century," *Sculpture Conservation. Preservation or Interference?*, ed. P. Lindley (Aldershot: Scolar Press, 1997) 195–208.
31. M. Gauthier, "Traiter la Ruine, ou le Visiteur?," *Faut-il Restaurer les Ruines?*, (Actes des Colloques de la Direction du Patrimoine.) Entretiens du Patrimoine (Paris: Picard, 1991) 72–73.

5

Minimal Intervention Revisited

Salvador Muñoz Viñas

The need for limits

On September 1778, Pietro Edwards was created Venice's Inspector for the Restoration of Public Paintings by the Venetian Senate. Following Edwards's suggestion, the Senate approved a set of rules (*capitoli*) establishing how these paintings should be restored. These rules were quite innovative at that time. For instance, they mandated that no restorer, 'even with the good intention of improving on the original, remove anything from the original, nor add anything of his own,'[1] that 'corrosive substances were not used that might endanger the untouched quality of the painting and corrode the paint layer,'[2] or that no materials 'which cannot be removed at will' should be used on the paintings.[3]

Pietro Edwards's *capitoli* can be seen both as the beginning of modern conservation (conservation as it is understood today) or as a brilliant, isolated exception: the very fact that someone had to explicitly forbid those practices suggests that paintings were routinely cleaned in an aggressive way, and that relevant portions of original works were often removed or covered by brushstrokes applied by the people in charge of their care. Nevertheless, time was on Edwards's side, and his views eventually became acceptable, and even taken for granted by many people.

However, in many ways, and throughout the many conservation specialties, conservators and spectators still seem to be concerned about excesses that are essentially similar to those that Edwards faced in eighteenth-century Venice. The criticisms in the controversy around the cleaning of Titian's *Bacchus and Ariadne*, in the debate over the Sistine Chapel frescoes, and in other polemics over the conservation of historical objects – such as the cleaning of Michelangelo's *David* in Florence,[4] the restoration of the Roman theatre of Sagunto in Spain[5] or the treatment of Barnett Newman's *Who's Afraid of Red, Yellow, and Blue III* in Amsterdam[6] – are not very different from the criticisms that are implicit in Pietro Edwards's rules.

The need to limit the excesses of conservation has been felt since its very inception. Most theoretical reflections on conservation deal with this topic in one way or another. Eugène Viollet-le-Duc reacted against the freedom with which historic buildings had been treated until the eighteenth century, which he thought to be excessive. For him, a damaged building should be necessarily restored to its original style, even if no evidence of the original state were available to the restorer: in these circumstances, 'the best option is to assume the role of the primitive architect, and imagine what he would have done.'[7] On the other hand, John Ruskin found that *any* restoration work was excessive, as he considered it to be detrimental for the authenticity of the original work.[8] Later on, the theorists of scientific or archaeological conservation (such as Gustavo Giovanonni or Camilo Boito) reacted against the excesses of restorations in the style of Viollet-le-Duc, which they saw as false recreations. With some nuances, these theorists suggested that missing parts should be restored to a previous state only, and only if there was enough evidence of that previous state to allow for a substantially *faithful-to-facts* restoration.[9] In very different ways, all conservation theorists were advocating some forms of 'restrained conservation.' All of them were saying that restoration could not go beyond some given point – even if that point was very different in each case.

In order to express these ideas, the limits were often defined as the 'minimal' action necessary to achieve some goal. For example, as early as 1904, the Sixth International Congress of Architects established that:

> Dead monuments [i.e. 'those belonging to a past civilization or serving obsolete purposes'] should be preserved only by such strengthening as is indispensable in order to prevent their falling into ruin.[10]

The *Carta italiana del restauro* (1932) also ruled:

> ...that in the case of monuments which are far from our uses and times, such as monuments from antiquity, any completion work should be routinely discarded; only the *anastilosis* should be considered – this is, the restoration of fragments with the addition of the minimal amount of neutral elements necessary to produce a coherent overall look, and to ensure good conservation conditions.[11]

And in a similar vein, Helmut Ruhemann resorted to this notion in order to establish some rules about conservation practice:

> Retouching, or 'inpainting' as the Americans aptly call it, should be kept to the minimum necessary to restore the coherence in composition and the character of a damaged painting.[12]

At some indeterminate moment in the second half of the twentieth century, however, the notion of 'minimal intervention,' which had traditionally been associated with the achievement of a goal, began to be used as an autonomous, self-referred concept. Soon, it became a popular term that was reproduced in books, articles and reports, and, eventually, it came to be considered as one of the most important principles of modern conservation. Chris Caple, for instance, described it as a notion that, along with 'reversibility' and 'true nature,' seeks 'to express a single quintessential guiding ethic for conservation,'[13] while for Caroline Villers, 'minimal intervention has been one of the dominant attitudes in conservation in the second half of the twentieth century.'[14] The principle of minimal intervention has even been described as 'the most important axiom in conservation.'[15] In this paper, the notion of minimal intervention will be analysed, in order to enable us to better understand it.

Minimal?

Language may fool us into thinking that 'minimal intervention' is an objective notion. 'Minimal' is an absolute term: it describes one extreme of a range of values. We are not dealing with those ambiguous, in-between notions or values that so often hinder precise communication. 'Minimal' does not mean 'rather little,' 'quite little,' or any other similar ambiguity. 'Minimal' is a much more exact concept. When we use the term 'minimal intervention' we may tend to think or feel that we are also expressing something that is an equally exact notion. However, this is an illusion. No intervention can be absolutely *minimal* and still be an intervention – both terms are inherently contradictory.

In fact, the term 'minimal intervention' can be considered to be an oxymoron. Just as the arrow in Zeno's paradox would never reach its target, it is always possible to imagine a conservation intervention that is, so to speak, a bit more minimalist. For example, bleaching a sheet of paper is not really a minimal intervention, as it could just be washed in water. Washing a piece of paper in water is not a truly minimal intervention as the sheet could be just gently cleaned with a soft eraser. Gently cleaning with a soft eraser is not really a minimal intervention, as the sheet could just be gently cleaned with an air spray. Gently cleaning the sheet with an air spray is not really a minimal intervention, as it could be *more* gently cleaned with an air spray by using lighter air pressure; and so on. Even changing its environmental conditions would imply an intervention which would affect the paper, increasing the number of hydrogen bonds, or reducing the distortions induced by inadequate RH conditions. The only truly minimal intervention would actually be leaving the sheet alone. In a practical (but also precise) sense, minimal intervention means no consequential intervention at all.

How minimal?

While the treatment of symptoms of some sicknesses can be objectively observed and even easily measured (e.g. increase in body temperature, changes in heart rate, raising or lowering of blood pressure), the effects of a 'non-minimal' conservation intervention can be neither observed nor measured in this way. Admittedly, we can measure the colour change of a painting caused by a particular cleaning technique and of an aged paper document after it has been bleached. However, while the standard temperature and blood pressure of a healthy individual are very well known, we cannot use any objective means to determine *if* the colours have undergone a non-minimal alteration: the colour change can be precisely evaluated, but there is no objective means to determine if the colour change of a restored painting or document is more intense than it should be.

When it comes to determining if a conservation treatment has been more or less extensive than it should have been, the criteria applied by a casual museum visitor looking at a painting may not be the same as the criteria applied by a historian engaged in doctoral research on art techniques and materials. The former might like to see vivid colours in the painting and a flat, taut canvas, while the researcher may prefer no historical or technical evidence to be removed. When observing a terracotta pot from the Bronze Age, the archaeologist's criteria may be quite different from those of a teenage student who is visiting an exhibition: the teenager is likely to prefer a fully restored pot with lively, captivating paintings, while the archaeologist may well prefer a very gently cleaned set of shards.

Then again, the opposite might also be true: the archaeologist may want the pot to be reconstructed in order to document the predominance of some shape in the pots of a particular time and place, while the young visitor might prefer to see the fragments 'as collected' from the site. Even if this were the case, it would not defeat the argument that different people may have different views, and that the determination of the 'right' amount of conservation is a subjective matter. No magnifying glass, no measuring tape, no ruler, no sensor, no sophisticated device with a complex name can tell us when and whether the conservation treatment we have chosen is 'enough.'

In spite of the absolute value implied in the word 'minimal' (or perhaps *because* of it), the notion of 'minimal intervention' cannot be of any help to us when it comes to determining the safe margins for any given conservation treatment. The acceptable minimum for a conservation intervention cannot be properly determined by any absolute principle: it is the result of subjective judgement. It depends on immeasurable (though indisputably real) factors: the tastes, preferences and expectations of the affected people. The 'minimal' cleaning of an aged varnish from a canvas painting, for instance, cannot be scientifically determined. If anything, science can tell us the morphology of certain layers within a painting or when a given layer was applied. However, the decision about removing certain layers of the painting while leaving others is completely subjective. Entirely removing *all* of an aged varnish is no less subjective than partially removing that same varnish or leaving it untouched. In all of these cases, the decision is the result of personal taste (usually the taste of empowered decision-makers). No scientific, objective measurement can justify any of these choices. These decisions can be supported with greater or lesser quantities of scientific data, and implemented with more or less scientifically-monitored techniques, but this does not make the *decision* to perform those operations any more objective.[16]

Thus, removing all of an aged varnish can disturb viewers regardless of the scientific data gathered before and during the process or of how delicately the intervention is performed, since they might prefer the object to have the particular appearance that aged varnishes produce. As Helmut Ruhemann realized while in Central America, some people may actually want some objects to remain in an aged condition:

> In 1956 I went to Guatemala on behalf of UNESCO for three months, to train three artist-craftsmen in picture restoration. The many religious pictures in the charming baroque churches badly needed attention, though few were of high value. Hardly any had ever been restored or cleaned, but when it came to removing the darkened varnishes from some of the sacred pictures I hesitated. It occurred to me that the population, the great majority of whom are *Indios*, descendant of the Maya in fact, might be appalled if they suddenly saw their Saints, whom they had always known with skin as brown as their own, emerge as white Europeans.[17]

Indeed, bringing an aged object closer to an as-new condition may be regarded as an excessive intervention by people who prefer an aged object to look aged. The conservation of El Greco's *Nobleman with his Hand on his Chest* was one of these cases. In 1996, this well-known painting was taken to the conservation laboratory of the El Prado museum, as it needed to be prepared for its temporary exhibition in Barcelona. The conservation treatment altered many features of the painting in a dramatic way: the characteristic black background had become grey thus rendering the shoulders of the nobleman visible; the golden sword handle was now silver-coloured; the artist's signature had disappeared; and the size of the painting had become smaller. Many people felt that the restoration had gone further than it should have, and the public uproar increased until it became what was described as a 'convulsión nacional' ('a national upheaval').[18] The conservator argued, however, that he had just returned the work to what all available evidence suggested was its original state.[19] Thus, the conservator and the public had very different ideas as to how minimal the intervention should have been – and neither the conservator nor the public could be proved right or wrong. The notion of minimal intervention cannot help in cases like this: as suggested above, if the notion is applied in a strict sense, all of these operations should be considered to be far from minimal; while if it is applied in a more vague, down-to-earth sense, there is no way to precisely determine what is *minimal*.

Intervention?

The Collins Dictionary defines intervention as 'the act of intervening,' while 'to intervene' is defined as 'to take a decisive or intrusive role.' In turn, 'intrusion' is defined both as 'an unwelcome visit,' and as 'the act or an instance of intruding' – 'to intrude' is 'to put forward or interpose (oneself, one's views, something) abruptly or without invitation.'[20]

Heritage objects suffer constant intrusions during the course of their lives: smoke, pollution, vandalism, and time, all of which modify these objects 'abruptly or without invitation.' Many of these modifications are considered to be positive or valuable (in this case we call them '*patina*'), while other ones are considered to be negative and undesirable (in this case, we call them 'damage').[21]

The conservator's work is also an intrusion in the object's life. Most conservation objects receive special care and attention, and no one would want a historical piece to be touched by anyone without special qualifications. Conservators are expected to be qualified professionals, and indeed they are the only experts who are normally allowed to physically intervene upon a vast number of historical objects: conservators are permitted to *alter* heritage. This alteration is supposed to be a desirable alteration, as it is expected to have positive effects upon the object itself. However, every conceivable conservation treatment has negative effects which, to some extent, defeat the theoretical tenets of the activity: whenever an object is treated, some of its original features are altered, some portions of the object's history are obliterated, and some information conveyed by the object is hidden or lost. Cleaning a Neolithic clay pot means that the analysis of the pollen deposited on its surface will no longer be possible; de-acidifying a paper sheet often alters its texture; lining a canvas painting hides the canvas texture and other evidence that could exist on the back of the canvas.

Even when new evidence is revealed through conservation, each modification implies the loss of historical evidence. For instance, removing the darkened varnish that obscures a paper map may render the map readable, but it will also remove information about how maps were used and cared for in the past; gluing together the fragments of a ceramic pot may allow the observer to better know the original shape of the piece, but the information contained in the broken edges (such as remnants of glue from previous attempts at repairing, or any information contained in the inner faces of the pieces) is hidden or lost; a cleaned painting may look better, but any information regarding the use and composition of old varnishes will also be lost. In summary, every possible intervention of the conservator has inherently negative consequences.

Fortunately, the intervention of the conservator may be 'abrupt or without invitation' but not only negative: the conservator's intrusion in the life of an object may also have positive consequences. In this sense, the principle of minimal intervention becomes somewhat confusing: why should the 'intervention' – the whole intervention – be ruled out, and not just its negative consequences? The principle refers to intervention as a whole, but why should the positive effects of an intervention be restricted? The principle of minimal intervention should only refer to the negative effects of an intervention (cancelling out bits of history, altering the original, reducing the number of possible readings of the object), but not the positive ones (increasing the lifespan of the object, preserving evidence, improving some preferred readings of the object).

If 'intervention' were understood as it is usually understood (as a set of actions with positive and negative effects) the principle of minimal intervention would be flawed, as it would rule out both the negative effects but also the positive ones. To cope with this contradiction, conservators unconsciously translate the meaning of the terms so that they can make sense of them – so that they mean what they are expected to mean. It is the same operation that has been described above in the case of the term 'minimal.'

It might be argued that the principle of minimal intervention is invoked because in real practice the positive and negative effects of the conservator's intervention cannot be separated. For example, when a stone statue is impregnated with a consolidation agent, some evidence is lost, as the pores of the stone are flooded with foreign matter; however, its lifespan may be increased; and if a painting is cleaned, evidence regarding the practice and materials used in the care of paintings in past times is lost; however, one of its possible readings, the painting as it was produced by the artist, is made much easier to view. Yet this does not defeat the point made above: the principle of minimal intervention does not actually mandate the minimalization of the *intervention* of the conservator, but just the minimization of some of its effects, namely its negative effects. It can thus be concluded that when we speak of 'minimal intervention,' we do not actually mean 'minimal intervention,' but a different thing altogether.

Beyond intervention

The term 'minimal intervention' may not be precise but, just as arrows do reach their targets in spite of any philosophical reflection, it must still be acknowledged that most people know what the notion of 'minimal intervention' means. Furthermore, the fact that it keeps being used demonstrates that it conveys an idea which, regardless of the accuracy of the words we use to convey it, is meaningful to people. In the following lines, the real meaning of this principle (i.e. the reasons why it is still useful) will be analysed.

With the exception of preventive conservation, conservation is all about altering objects. Restoration alters an object, and it does so in a perceivable way. Observers can tell if a painting has been restored because the tone of its colours has changed, its surface is more flat, or its missing parts are less discernible. In these cases, the condition of the object itself reveals that it has been restored. Indeed, it has been *deliberately* modified. The whole point of restoration is to change an object in a way that can be noticed by the observer, who will hopefully obtain some benefit from the alteration (for example, increased aesthetic pleasure, increased local or national pride, or increased knowledge).

However, conservation (conservation as opposed to restoration) is not intended to be noticed by the observer. For instance, resizing a paper sheet or consolidating an old wooden frame with a synthetic resin is a typical conservation treatment that results in an altered object: at the very least, these fore-mentioned treatments modify both the mechanical properties and the material composition of both objects. These modifications, however, are not easily discernible by a regular observer. While restoration treatments are deliberately noticeable, non-restorative conservation treatments are always as least discernible as technically possible.

Conservation and restoration often happen together. For example, lining a canvas for reinforcement often results in flattening, and washing a sheet of paper to remove some acidic degradation by-products often leads to it whitening. Even though some of these effects may or may not be intentional, it does not preclude the idea that conservation does change the object. Admittedly, restoration deliberately aims at changing noticeable features of the object, while conservation aims at changing unnoticeable features, but at the end of the day all these activities are about changing some of the object's features.

The changes introduced by conservation are ethically relevant as far as they affect the object's meanings. Restoration is carried out to allow an object to convey some particular meanings more effectively and/or for a longer period of time. For instance, a certain amount of pigments embedded in a calcium carbonate layer, with paint, smoke and aged varnish on its surface – an aged fresco, in other words – may be made to reveal not its own history, but the work of a long-dead painter. This can be achieved by depriving it of as many of the imprints of history as possible (such as smoke and aged varnishes), so that it no longer looks like an aged object, but rather like the recently-produced work of the artist. This alteration means that the object may better represent the work of an artist. Unfortunately, it also means that that same object may no longer convey other possible meanings, such as the age of the painting, former historical events, or the work of other painters that might have updated or retouched the painting at some given moment in time. Thus, an ethical dilemma is posed to the conservator: Should revealing the work of the artist be the main goal of the treatment? Or should it be preserving as many traces as possible of the actual history of the painting? This kind of dilemma (the choice about which meaning should prevail and which ones should be sacrificed) lies at the heart of nearly every conservation controversy.

In conservation (conservation as opposed to restoration) the most obvious meanings are not altered, while non-obvious features (tensile strength of a canvas, pH of a paper, rigidity of a wooden frame, degradation by-products on a metal coin) may be dramatically altered. These features are usually meaningless for most people, and this is why restoration treatments account for most of the controversies in the conservation world. However, these less-obvious features can tell very interesting stories for people who can interpret them – specialists, such as chemists, historians, and archaeologists, among others. Only those with specialized training may be aware of the value of the cleavage of the particles of a clay fragment at a microscopic level, or the scientific relevance of a black deposit on a marble statue. Furthermore, no one can actually tell what information may possibly be obtained from *any* feature of the object by means of a presently unknown technology that could perhaps be developed in the future. It has even been suggested that the DNA of great personalities from the past could be obtained from small remnants of organic materials that could still exist within very small cavities in the objects that they produced or touched.[22]

Modification is what conservation is all about, and some meanings of the object will necessarily be sacrificed in the process. The principle of minimal intervention can be better understood in this light. When we speak of 'minimal intervention' we do not actually care about the intervention itself. Rather, we care about the losses in the meaning-bearing ability of the objects that any modification may cause to the object: it is the decreased ability of the object to transmit some messages, whether social, ideological, aesthetic or scientific, that we really care about. The principle of 'minimal intervention' does not actually refer to the intervention as such, but to the loss of potential meanings of the object. It is not actually a principle of 'minimal intervention' but rather a principle of 'minimal loss of potential meanings.'

From 'minimal intervention' to 'balanced meaning-loss'

However, the criticisms described in the preceding section of this text remain valid even if the principle of 'minimal intervention' is interpreted as the principle of 'minimal meaning loss.' For instance, just as it is always possible to find a 'more minimal' intervention, it is always possible to find a 'more minimal' meaning loss – unless we decide to carry out on the object the truly 'most minimal intervention:' no modification at all. Even if the notion of 'meaning loss' is substituted for the notion of 'intervention,' 'minimal' still remains an improper adjective. So, what does 'minimal' really mean in this context?

The conservator does modify the object, as do other people (owners, vandals, archivists, priests, readers, archaeologists). The conservator, however, endeavours to adapt the object to the tastes and needs of as many observers as possible, be they laymen or experts. However, the needs and tastes of different observers may be extremely varied. Some scientists may be interested in specific trace elements contained in surface deposits on a marble statue, an art historian might be interested in knowing one of the successive states of the statue, and a casual observer might well prefer the statue to be cleaned, polished and well presented. When making a conservation decision, it may not be possible to satisfy all of these 'tastes.'

However, even if there were a general agreement over which meaning should prevail, things may not be so simple, since the 'observers' include not just contemporary observers, but also future observers, who could have tastes and needs that we cannot foresee. And since future observers also need to be catered for (one of the main purposes of conservation is to increase the lifespan of the object), a new requirement is added to an already complex task.

Conservation principles help conservators in the decision-making process by recalling some of the least obvious things that need to be taken into account. 'Minimal intervention' (or 'minimal meaning loss' if so preferred) is one of these principles: it reminds the conservator that each change introduced in the object might compromise meanings that future observers may consider to be valuable, or may diminish its value as scientific evidence. 'Minimal intervention' is a *caveat* against the temptation to modify the object according to some particular contemporary tastes and needs while forgetting that other possible meanings can be impaired or lost in the process, and that these meanings could be valuable for other people.

One of the most concise reflections on the principle of 'minimal intervention' was written by Chris Caple: 'The problem with minimum intervention is that it is not a complete statement. Minimum intervention to achieve what?'[23] This is a brilliant way to express the relativity of this principle, especially when we are aware that the goal of any intervention is to preserve or strengthen some meanings of the object. Certainly, the 'minimal intervention' that this principle calls for depends on the goals of this intervention – or, in other words, on the meanings we want the object to convey. The minimal intervention necessary to restore a painting's original appearance may be very different from the minimal intervention necessary to prevent further deterioration of that same painting; the minimal intervention necessary to make a historical document look flat and clean may be very different from the minimal intervention necessary to keep its text readable; the minimal intervention necessary to make a sculpture suitable for religious ceremony may be very different from the minimal intervention necessary to preserve its documentary value.

Caroline Villers introduced the expression 'post minimal intervention' in order to express the need for a conservation principle which could replace the principle of 'minimal intervention,' thus overcoming its limitations.[24] Many of these limitations derive from the imprecise choice of terms: neither 'minimal' nor 'intervention' mean what these words usually mean. The principle does not call for any minimal intervention. Instead, it actually means something different and more involved: it mandates that *conservation should enhance or preserve the preferred meanings of the object while impairing as little as possible its ability to convey any other meanings*.

In the expression 'minimal intervention,' the term 'meaning loss' could be advantageously substituted for 'intervention,' but there are also many other adjectives that could replace the term 'minimal.' 'Balanced' would be a good candidate: a notion such as 'balanced meaning-loss' reflects the need to achieve an equilibrium between gains and losses of meanings, which is the rationale behind the principle of minimal intervention. Also, it suggests the idea that conservation objects can have different meanings.

Conservators, or conservation decision-makers, need to find a happy medium between preserving each and every feature of the object (i.e. all of its potential meanings), and its free and complete alteration to the contemporary observer's tastes or needs (i.e. its complete alteration to make a particular reading prevail). Sometimes, this sweet spot is easy to find (for instance, when the conservation treatment is unanimously expected to preserve most of the evidential features of the object). However, sometimes this is much more difficult to determine (for instance, when the object possesses different meanings for different people). Unfortunately, there is no rule that can make these decisions simpler. In fact, by reminding the conservator of the negative consequences of the intervention, i.e. by introducing more factors in the decision-making process, the principle of minimal intervention (or balanced meaning-loss, or whatever we call it), can actually make things more complex. More complex, admittedly, but ultimately better: more satisfying, for more people.

Notes

1. 'Che alcun professore neppure con buona intenzione di migliorar l'opera levi cosa alcuna dall'originale o vi aggiunga qualche parte di proprio.' A. Conti, *Storia del restauro e della conservazione delle opere d'arte*. (Milan: Electa, 1988) 166; the English translation in notes 1, 2 and 3 are taken from A. Conti, *History of the Restoration and Conservation of Works of Art* (Oxford: Butterworth-Heinemann, 2007) 191–193.
2. 'Che per ispedire solecitamente il lavoro non si adoperino corrosivi capaci di togliere la virginità del dipinto e bruciare il coloure.' (Conti, *op cit.*, 165).
3. '[the restorers] si impegnano di non usare sui quadri ingredienti che non si possano più levare, ma ogni cosa necessariamente adoperata sarà amovibile da quelli dell'arte ogni volta si voglia.' (Conti, *op cit.*, 165).
4. To celebrate the 500th anniversary of Michelangelo's *David*, the Florentine authorities decided that the statue should undergo a gentle conservation treatment. The specialist Agnese Parronchi was designated the task and subsequently decided that the best cleaning method would be a dry treatment using chamois cloths. However, the scientific committee considered that wet poultices would be preferable, as these would more thoroughly remove surface gypsum deposits. This disagreement soon entered into the realm of public controversy, and resulted in the resignation of Parronchi. A. Riding, "Question for 'David' at 500: is he ready for makeover?," *The New York Times*, 15 Jul. 2003 (available at http://query.nytimes.com/gst/fullpage.html?res=9A0DE0D7173CF936A25754C0A9659C8B63), accessed on February 6, 2008.

5. The ruins of the Roman theatre of Sagunto were deemed for restoration in the eighties and the architects Giorgio Grassi and Manuel Portaceli were entrusted with the job. From the start their project was publicly controversial, as it proposed to cover the original remains. However, the restoration went ahead and an entirely new concrete building was erected over the remnants. The new building supposedly imitates an idealized Roman theatre. The works ended in 1993, but the controversy was far from over. After an arduous legal battle, the restoration was declared then finally confirmed illegal (2007). H. Esteban and P. Salazar, "El Consell duda entre levantar el mármol antes de elecciones o eludir la sentencia," *Las Provincias*, 4 January 2008: 2–3.
6. *Who's Afraid of Red, Yellow and Blue III* is an abstract painting by Barnett Newman (Stedelijk, Amsterdam). In 1986, it was vandalized by a madman, who slashed the full extent of the red monochromed surface. Daniel Goldreyer, a New York-based conservator, was commissioned to repair the damage. When the painting was sent back to the museum, its new appearance attracted attention and criticism by certain art historians and scientists. As it was discovered, Goldreyer had repainted the entire red surface area with a paint roller. This contentious discovery caused public uproar, due to excessiveness; however the restoration remains unchanged. M. Johnson, "Was a Masterpiece Murdered?," *Time*, 30 December 1991: 15.
7. '. . . le mieux est de se metre a la place de l'architecte primitif et de supposer ce qu'il ferait.' E. Viollet-Le-Duc, *Dictionnaire raisonné de l'architecture française du XIe au XVIe siècle* (Paris: Librairies-Imprimeries Réunies, s.f.), 32. Translation is mine.
8. See, for instance, *The Seven Lamps of Architecture*, VI, 17–20.
9. See, for instance, I. González-Varas, *Conservación de Bienes Culturales. Teoría, historia, principios y normas* (Madrid: Cátedra, 1999) 227–38; J. Jokilehto, *A History of Architectural Conservation* (PhD Thesis, 1986) 18.3. (http://www.iccrom.org/eng/02info_en/02_04pdf-pubs_en/ICCROM_doc05_HistoryofConservation.pdf), accessed on September 26, 2007.
10. Sixth International Congress of Architects, *Recommendations of the Madrid Conference* (1904) 'The preservation and Restoration of Architectural Monuments. 2nd recommendation' (http://www.getty.edu/conservation/research_resources/charters/charter01.html), accessed on September 26, 2007.
11. 'Che nei monumenti lontani ormai dai nostri usi e dalla nostra civiltà, come sono i monumenti antichi, debba ordinariamente escludersi ogni completamento, e solo sia da considerarse la anastilosi, cioè la ricomposizione di esistenti parti smembrate con l'aggiunta eventuale di quegli elementi neutri che rappresentino il mimimo necesario per integrare la linea e assicurare le condizioni di conservazione.' Consiglio Superiore per Le Antichità e Belle Arti, *Norme per il restauro dei monumenti* (1932) Art. 3 (http://www.webalice.it/inforestauro/carta_1932.htm), accessed on September 26, 2007. My translation.
12. H. Ruhemann, *The Cleaning of Paintings. Problems and Potentialities* (New York: Haecker Art Books, 1982; first published in 1968) 241.
13. C. Caple, *Conservation Skills. Judgement, Methods and Decision Making* (London: Routledge, 2000) 61.
14. C. Villers, "Post minimal intervention," *The Conservator*, Volume 28 (2004): 3.

15. 'The most important axiom of conservation is: Minimal intervention is best' (The Vertebrate Paleontology Department Of The Florida Museum Of Natural History, 'Fossil Preparation and Conservation,' in (http://www.flmnh.ufl.edu/natsci/vertpaleo/resources/prep.htm) accessed on April 9, 2007).
16. See, for instance, D. Lowenthal, *The Past is a Foreign Country* (Cambridge: Cambridge University Press, 1985); S. Muñoz Viñas, *Contemporary Theory of Conservation* (Oxford: Elsevier/Butterworth-Heinemann, 2004); D.E. Cosgrove, "Should We Take It All So Seriously? Culture, Conservation and Meaning in the Contemporary World," *Durability and Change. The Science, Responsibility and Cost of Sustaining Cultural Heritage*, eds. W.E. Krumbein, et al. (Chichester: John Wiley and Sons, 1994) 259–266.
17. Ruhemann, p. 46.
18. J. Luzán, "Las heridas de El Greco," *Tiempo*, Volume 888 (1999): 35–43.
19. R. Alonso, "En defensa de una restauración," *El Mundo*, 8 June 1999: 60.
20. *Collins Dictionary of the the English Language. Second Edition* (London & Glasgow: Collins Sons & Co., 1986).
21. J. Ashley-Smith, *Risk Assessment for Object Conservation* (Oxford: Butterworth-Heinemann, 1999) 99–119; J. Ashley-Smith, 'Definitions of Damage,' (http://www.palimpsest.stanford.edu/byauth/ashley-smith/damage.html), accessed on March 30, 2007.
22. J. Beck, "What does 'Clean' Mean?" *Newsday*, 6 October 2002: A26.
23. C. Caple, *Conservation Skills. Judgement, Method and Decision Making* (London: Routledge, 2000) 65.
24. Villers, p. 3.

6

Practical Ethics v2.0

Jonathan Kemp

Introduction

The title of this chapter alludes to Peter Singer's 1979 book, *Practical Ethics*.[1,2] Singer's version of applied ethics (in his case, preferential ethics) has antecedents in Jeremy Bentham's Utilitarianism, roughly defined as being that an action is good if it creates the greatest happiness for the greatest number. Bentham conceives of an algorithmically based calculus for judging whether an action is good or not. However, the calculus can only work in a commensurable and closed system, that is, in a state of being self-defined and isolated from environmental influence (as in, for example, a dictionary, which is a closed system in that it defines each word in terms of other words, all of which can also be found in the same dictionary). Thus environmental influences, *vis à vis* incommensurability, present challenges to any ethical theory that contends that the right thing to do is the action that promotes the most overall good. Any disagreement about the commensurability of values, as, for example, in contemporary notions of 'stakeholder consensus' or 'interpretative community,' means that any Bentham-like utilitarian calculus is not even theoretically possible.

Thus, as codes of ethics can only be successfully applied in a closed system and, as with many human agencies, conservation generally operates to a lesser or greater degree in open-ended systems, then all conservation actions are bound to fail when measured against any one version of the ethical codes of conservation. This chapter considers the execution of ethical behaviour in both present and future settings with the offer of a beginning to the resolution of this inherent contradiction.

Terminal beach[3]

With many artifacts there is a pronounced instability in identifying particular components as sites of authenticity in the sense of 'original material,' traditionally one aspect of an object charged by the assignation of a 'truth-value' that legitimizes some aesthetic experiences. It is because of this particular notion of authenticity that conservation has lain down its principles on the bedrock of scientific analysis, outwardly assimilating those methodologies into its own ethical codes and practice.

One consequence is that attributions of authenticity are always open to modulation by the development and availability of this or that scientific technique to narrow down probabilistic error.[4] As the methodologies of material science have become the most authoritative means of object description then, concomitantly, they have also become the authority for legitimizing much of conservation practice. And in so doing they ensure their own transmission, and that of their essentialist epistemology, through the preservation of the material object.[5]

Flip[6]

What does authenticity have to do with ethics?

Attempts have been made to pull this predominant focus of conservation away from its perceivably narrow concentration on the material condition of an object. Notions of an object being an actor within a social network and conservation as a social process engaged in the production of cultural objects have been discussed as framing any concern with an object's material safeguard and presentation. However, from within conservation the narrower focus remains dominant in binding authenticity to ethics, as summed up by Frank Matero: 'Implicit is the notion of cultural heritage as a physical resource that is valuable and irreplaceable – an inheritance that promotes cultural continuity.'[7]

So conservation practice is well rooted in privileging the retention of the physical integrity of the object through the minimizing of loss. Any restorations are viewed as potential un-tetherings of the object from a state of authenticity, in intent or condition, and a state against which any difference, ultimately, can be quantified scientifically. If inauthentic expressions (viz. restorations) somehow deceive, then additions to compensate for loss cannot, under this rubric, be original or authentic. This has led to the current practice to cue additions, where they are necessary, so as to be discovered as inauthentic, and thus to allow people to have an impression of the object *qua. object* but to become aware of just where and what is unoriginal.

This focus on material authenticity underpins the preferred notion of disclosure embodied in current codes of ethics: that which is altered is documented in the object itself and is detailed in the accompanying record.[8] Thus, when a work is addressed, its condition of authenticity can be evaluated through the legacy of a practical 'instruction manual' constituted by its documentary record (where it exists) and in its own physical record when its component-on-component relations are decipherable and understood within the then prevailing modes of practice and codes of ethics.

Subsequently, when the work is next addressed, the process of conservation then privileges a particular version of its authenticity instantiated by the dominant zeitgeist as described in any code of ethics then circulating, whilst the instructions as embodied in this manual are interpreted. This assumes, of course, that codes can be rigorously applied in a closed system, and that the physical notation and material conventions of the work can be understood by successive generations of conservator-restorer-scientists.

Flop

So are things simply authentic or not?

A common model of conservation is represented here as a triadic graph[9] (Figure 6.1).

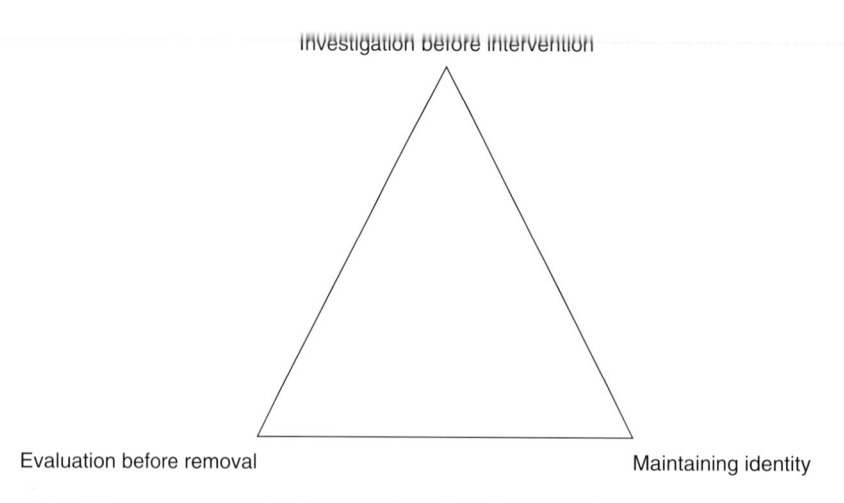

Figure 6.1 *Diagram representing the core ideas that determine Conservation Actions.*

This diagram fairly well represents the dynamics involved in designing and executing conservation actions, with each corner more or less representing a core idea in current conservation methodology:

- **Investigation before intervention**: the injunction to perform research on, and documentation of, all relevant evidence before and after any intervention.
- **Evaluation before removal**: the injunction to respect the process of history in its cumulative record of activity reflected in the object and identified as denoting varying cultural beliefs, values, materials, and techniques executed over time.
- **Maintaining identity**: the injunction to safeguard authenticity – herein understood as an epistemologically relative term associated with the material and material process in an object and its authorship or intention – co-joined with the obligation to execute minimal physical intervention to help re-establish structural and aesthetic legibility and meaning whilst allowing future treatment options.

Put different kinds of objects in the frame and it becomes a pretty rugged plane of (in)consistency, with the idea of material authenticity tugged at from all sides. For example (though perhaps arguably), the objects of ethnographic conservators will tend to bunch around the lower and left side of the triangle, as their material changes are seen as the attrition of an authentic history, whereas those of conservators working in design museums tend to cluster more to the right side, as a 'return' to some sense of an original state is the curatorial objective. And where some conservation practice seeks to maintain the current condition of an object – or at least the illusion of it – other work, for example, some architectural conservation, seeks to establish continuity through controlled alteration, thereby spreading itself into the lower right of the triangle.

So while this triad emphasizes the dynamics within conservation that attempt to maintain and transmit cultural continuity through the protection of valuable physical resources, it also highlights the problems in understanding just what material authenticity is. It is also a step along the way in exposing ethical codes as being the products of social processes mediated by a technologically based practice, with add-on values that accord with a particular custodial community's goals, as indicated above.

Lotta Continua[10]

Even if this contemporary notion of authenticity accepts the vagaries of identifying sites of original material it can still seem to call upon supporting variations on the theme of 'original intention,' and so pushes along with an essentialist model of cultural production. By this I mean that if the organization of a particular community determines the form of ideas held by the people within it, then, under the current epistemological landscape, conservation can be described as a compact social network which internalizes its values and social arrangements in collectivized representations which are thereafter treated as, in effect, essences.

As it is, codes of ethics are intended to produce agreed behaviours. Within conservation they do so not by invoking clearly defined goals, rather by providing aspirational guidelines in treatment decision-making that reflect, albeit perhaps by consensus, the guiding philosophy of the conservation constituency. As such, and without explicitly formulating it, these codes progress with an *either/or* polarity around the notions of authenticity and truth, whether in material or intention, with, for example, prompts as to what kinds of intervention are considered right or wrong (so thereby placing restoration on the pillory of reversibility).[11]

A question of re-entry[12]

The advent of variable media in contemporary cultural production is one development that has called into question both those values and the methodologies by which custodial groups preserve, care for, and redisplay cultural artifacts. In particular, new media art has detoured notions of authorship, intention and material authenticity by making vigorous explorations in collaborative working, by way of variously plugging into interactivity, randomness, networking, and virtuality, and by mining notions of open code, open hardware, and open documentation in its own technical development. Whilst new media art might suffer from technological obsolescence (one concern of digital media preservationists[13]), these aspects, centred around hybridized, contextually-based, live or time-based productions, represent a snowballing defiance of traditional conservation methodologies rooted in essentialist-based concepts of viewer, artist, and the art object.[14]

So going back to the question, *are things simply authentic or not?*, it can be reframed by hypothetically plotting a work at any given time between three temporal axes where each axis nominally describes variables emanating from the impossible-to-return-to ground zero of an object's origin. The z-axis plots any significant change to an object's function, the y-axis any change in how the object is interpreted, and the x-axis plots any change in original material (Figure 6.2).

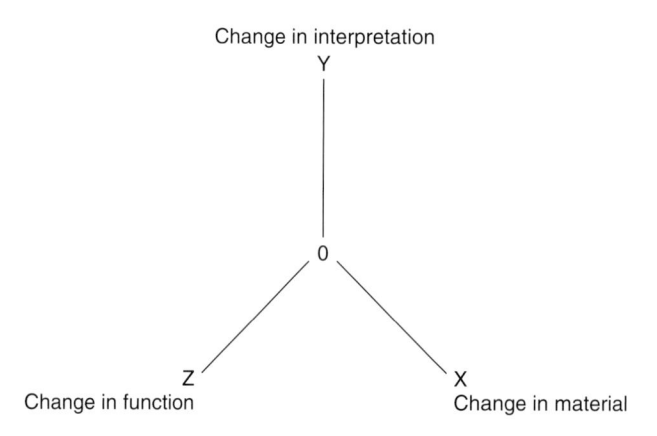

Figure 6.2 *Some variables used in determining an object's authenticity.*

Without getting bogged down in defining other relativistic states that could also emanate from ground zero, other axes worth mentioning might include changes in design and authorship. By playing around with this thought experiment it becomes more apparent that objects don't fit into *either/or* categories of being authentic or non-authentic when plotted along the given axes, and that changes along multiple axes will give each object a unique topology, with its edge nearer or further away from its ground zero. For example, a panel of stained glass described as medieval tends to comprise of little original glass, still less original lead – as restoration, namely the return to a design that is known, has been a regular conservation process until at least the 1990s – yet can still be described as being authentic. If plotted schematically along the various axes, following the thought experiment outlined above, and with its co-ordinates joined up as an outline (its topological edge), then this shape is going to be pretty far away from the ground zero point of the panel's origin, especially when this outline is compared with one, say, drawn for a gravestone that has remained pretty much untouched in its original setting.

Furthermore, few objects will have the same co-ordinates at any given time in their history, and museum objects can never maintain the same co-ordinates. They may, however, be nearer their zero point if they were somehow made as part of the museum, as in the case, for example, of Frank Moody's late 1860s ceramic staircase at the Victoria and Albert Museum, London. But even objects in their original context will change as an object deteriorates, or is re-used in some way by a subsequent user group; and within a collection the co-ordinates of an object invariably change whenever it is conserved or redisplayed.

The point of this thought experiment is to show that any sense of authenticity, loosely pinned to this schematic, is always going to be a ride along a trajectory from which, at any one point, the object will have stronger or weaker genealogical links to its origins. And once this notion of authenticity as being 'vectorized' is established and the care of an object is framed in this way, it becomes more apparent that the preferences of conservators, curators and others invariably alter the co-ordinates (and topology) at any given time. This model also seems to take on some of the aspects described by new media art, so that it begins to appear that any work exists as something like a collaborative production, only on a longer and more drawn-out timescale. Pip Laurenson, Head of Time-based Media Conservation at Tate, London, writes that she '. . . would suggest that the concept of authenticity operating in the traditional conceptual framework of conservation is appropriate for a framework in which the objects of conservation are the autographic arts but inadequate for works which are not.'[15]

The underlying suggestion in this chapter is that the traditional concept of authenticity as described by Laurenson is inappropriate for *any* work, with part of the thesis being an attempt to show that *all* autographic works actually have an allographic component as when any one work is considered at two points in its history, each iteration's qualities will necessarily be different to the other, yet each will still be considered as 'the work.'

This reframing is intended to shift any notion of assigning truth-value away from the difficult condition of material authenticity, and onto documentary notation, as authenticity becomes a matter of the (play of) accuracy with which the present cultural apparatus plots an object and provides a full commentary on how its particular interpretation relates to that of its predecessors. Another part of the thesis in this chapter calls for a methodical documentation of such cultural schemata as part of what this author sees as the necessary unveiling of cultural production.

Plot construction[16]

What does such a plotting really mean?

As discussed earlier, as a part of conservation's concern with methodological efficiency the profession has subscribed to some particular forms of disclosure nominally centred round clear documentation. In the current preferred version (of disclosure) such documentary features are designed to pin down decision-making by conservators onto the bedrock of empirical evidence, so that, for example, future custodians can reverse-engineer the present, both from the 'then' state of the object and its treatment record. So usually this documentation is where the immediate facts relevant to the object's care are recorded for (a variable) transmission amongst a close-knit community of custodians and other, sometimes undisclosed, 'stakeholders.' However, the wider interpretative, cultural, economic and political contexts for decision-making that directly affect object care are generally absent from any of these transmissions. Sometimes, and off somewhere else, they might cause a lot of localized noise, but noise from which it is pretty impossible to modulate a clear signal for the forensic reconstruction of their content.

An example helps illustrate this latter point: A monument in a museum's collection cannot be removed before the redesign and renovation of the internal architectural space surrounding it begins. Sometimes this can result in the removal of material from that object to accommodate the intervention of the new construction. The material loss will be evident in the object itself, and described in the conservation record, but invariably the policy decisions that effected the alteration or loss, or any localized opinion raised for or against those policies, will not or cannot be transmitted in any object record.

Confront the essentialist rubric of any code of ethics with something like this and attention is immediately focussed on the epistemic relativism inherent between what a code specifies and the contingencies in any execution.

But contrast this against an ideological desire to fully disclose the contexts of decision-making that shape an object's current status (another example of this might be why some restoration has not been removed), and a rich heuristic is provided for the entry of any archive as an asset into the wider knowledge economy, and conservation's longstanding commitment to disclosure positively mandates its entry into this changing economy.

Radio-On[17]

Section 9.4 of the UK Museums Association's Code of Ethics for Museums (2002) urges that museums should 'develop mechanisms that encourage people to research collections, develop their own ideas about them and participate in a variety of ways in shaping the interpretations offered by the museum.'

Globally speaking this injunction is beginning to be addressed across various institutions with the development of collection-management tools for the tracking, archiving and interrogating of object documentation. As such, digital technological systems have thus enabled institutions to rethink the relationship between their information holdings and its accessibility to the public. Recent discussions in New York (2006) and London (2007) have focussed on furthering the sharing and interoperability of conservation information and platforms, both within a particular institution and in exchanges between many-to-many institutions, along with the possibilities of that wider public access.[18]

However, the extent of public access to such documentation has been seen by most to be different in kind from that of any inter-institutional exchange. The debate is currently labouring around questions of how and when such information might be shared with the public, with issues of intellectual ownership cited, including 'the risk of misinterpretation or misuse of raw, uninterpreted data … especially as they relate to proprietary authorship and to works in progress that are destined for publication but not yet adequately advanced for dissemination.'[19] Furthermore, institutional sensitivities regarding treatment policies and histories have also been raised in support of a tiered (or gatekeeper) approach to access.

Many worlds[20]

So what has this got to do with conservation ethics?

Earlier, I indicated how I thought that the notion of truth-value might be re-inscribed into documentary notation. Such a re-inscription has the potential for the creation of information free from the historical hostage taking that has traditionally reflected the privileges of the dominant cultural powers in its ordering of information into categories of intellectual property.

To transfer any sort of knowledge (especially for the benefit of those to come) it is apparent that it has to be encoded into a medium that will allow it to be transmitted and decoded successfully. Thus, the technical means of description and transmission available readily limit the scope of the transmissions permissible, and impact on the type and extent of the economy to be managed.

Slides in drawers, files in cabinets, desktops and offices (propagated in the iconography of current proprietary computer operating systems), all favour a heavily biased model of scholarly knowledge management, one of restricted levels of access and privilege centralized around a gatekeeper model of one-to-many exchange.

Technology defines practice that in turn creates theory. So in the wider context of digital technologies, the gatekeeper economy certainly begins to appear too narrow and proprietorial, with its continuing focus on the interpretative control of information. The last decades have seen the transformation of knowledge-behaviours by digital technologies including the Internet, with keys to this rapid shift being both the advent of the free software (or 'open source') movement and Web.2.0. This shift has seen the production and distribution of knowledge moving away from being a flat one-to-many model to a model of many-to-many content-generation. This, in turn, has led to the creation of an information-rich economy in which the circulation of knowledge and exchange in cultural content is the norm.

In the United Kingdom, especially for public bodies, this economy has been at least nominally underwritten by recent legislation with the UK's Freedom of Information Act (2000), the terms of which came into full force in January 2005. In essence, the Act gave the right of access to documents and information held by public institutions. There are a number of caveats to this right (national security, commercially sensitive material, possible infringements of intellectual property rights and so on), but the kind of information covered by the Act includes all conservation records, with the right for the public to be given the information in the form in which the public institution holds it. 'If an enquirer is dissatisfied with an institution's response to an enquiry, the person has recourse to an independent commissioner for information, who is empowered to adjudicate.'[21]

In short, the digital economy is the most important ideological tool for scholarship since the printing press, and all information that can be online should be online, because that is the most efficient way to distribute material to the widest possible audience. The digital economy, including the Internet, is not just an adjunct to an existing environment; rather, it is the new environment, and looking for ways to gate the force of its distributive efficiency exposes proprietorial and territorially entrenched behaviours. Any debates around what caveats might or might not apply, whatever their merits, forgo the consequences of this economy, the principle of which manifests in the production of a cultural framing of technology in an ethic of practical benevolence. This means that, in the domain of conservation, the significance of instrumental reason is fully recognized, but tempered by the notion that human agency is not constituted solely by a disengaged rationality operating consistently or in a closed system.

Back to the future[22]

Thus, true openness of an object's documentary record might be through exposing its different versioning in any implemented content management system. Contributors use verifiable identities, and there is some moderation of the record by, for example, the expert oversight of everyday contributors, with an institutionally authorized versioning including digitized records closed to real-time editing.

Version control (also known as revision control), is the management of multiple revisions of the same piece of information. It is most commonly used in engineering and software development and other areas where information content may be worked on by a team of people, typically blueprints, electronic components, and managing successive developments in a software application's source code.

Changes to version-controlled documents are usually identified by incrementing an associated number or letter code, the 'version number,' or, as is the case with the wiki on which this chapter is being written (a wiki is a collaborative website whose content can be edited by anyone who has access to it), a simple 'revision' number and, as an option, associated historically with the person making the change. Just now I'm logged as working on 'Revision 202 . . . 2008-03-28 13:49 UTC by jk.' And now I'm logged on as 'Revision 236 . . . 2008-04-18 15:49 UTC by i-83-67-116-113.freedom2surf.net,' as I'm working from a different computer on a different day. Anyone can compare the differences between the two versions by hitting 'View revisions' to 'rollback' to the relevant version numbers by using the 'compare' function featured on this particular wiki engine.

The most widespread example of this form of knowledge management is Wikipedia. Wikipedia is the bastard child of a failed attempt at providing a free online encyclopaedia where anyone could submit content that would be reviewed in a seven-stage process by expert editors. It was a project born out of the politics of encouragement of openness and extreme decentralization. Its model of collaborative entries is founded on the belief that entries would 'self-edit' in a series of redrafts where someone who knew more about one part would edit the original entry, while someone else would make any grammatical corrections necessary, and so on by way of interdisciplinary and creative conversations.

Several challenges are immediately apparent as, for example, the transfer of the implicitly held knowledge of individuals is not usually available. On the other hand, explicit documentation in the form of written reports, database entries and images require a heuristic from which they can be edited to maintain an optimal signal-to-noise ratio; that is, how well a receiver can recover the information-carrying signal from the transmitted version and hence how reliably information can be communicated. The British Museum's Institutional Summary for the 2006 Mellon Report states that, 'as with enquiries from other museum users, the knowledge that records may be scrutinized has the added advantage that it increases professional accountability and responsibility, and leads to improved standards of documentation.'[23]

Another challenge is that the medium used for information-storage must itself be quickly accessible and replicable, and there must be a successful transmission of its own means of production; i.e. the magnitude of the signal-to-noise ratio inherent in a communication system must be factored by the inclusion of its own manual in its transmissions.

Ogres and onions[24]

The open-ended offer of this chapter is groundwork towards defining a kit with which any user can reconstruct the sort of decision-making instruction manual used at the particular time of the object's revision, a descriptive specification sheet derived according to whom is reconstructing the available components.

From the closed viewpoint, for those guarding information, it is the world of the slide library, not a library for the people, but the fiefdom of ownership. But in the contemporary networked landscape it is becoming readily apparent that closed knowledge loops simply cannot remain sustainable.[25] Within the context of necessarily decentralized groupings or communities, information could readily have quite different meanings attached. And from this open access angle, it is all about emergence, community and subscription to a new model of knowledge, with primarily textual solutions to ethical dilemmas as the order of the day.

Such an ethical approach is thus descriptive: the decision mechanisms and social processes through which, for example, a museum is produced are tied more closely to the care of its objects. For example, the degree of deviation from the normative (*qua* scientific) methodologies subscribed at a particular time in that object's care is openly indicated as part of the process of this disclosure (as a varied account of disengaged rationality!).

This ethic is also moderately prescriptive because its methodologies are thus to be adopted as the prevailing ethos for object care across that museum. This in turn means that the museum rigorously adopts those methodologies that it at any one time subscribes to in governing all of its technical operations. This approach also introduces commensurability, a level of agreement applicable to *all* involved in the technical care of objects, as the imperative for a preferential or utilitarian ethic to be applied successfully.

Such a conception helps for a more systematic plotting of the object along the hypothetical axes of authenticity introduced in this paper, and this reinforces the understanding that the current version of an object (or asset) is a part of its continuing history.

In summary, I suggest that an understanding of any fault-lines between the application of conservation's codes of ethics and its actions, and between the material authenticity of an object and authenticity of the observer's experience, can be neatly rounded out in the object record, a record that should become a major part of any institution's current knowledge economy, as well as a systematic transmission to the future.

Notes

1. P. Singer, *Practical Ethics*, 2nd edn (Cambridge: Cambridge University Press, 1979) 1993.
2. The title of this chapter, '*Practical Ethics v2.0,*' was also chosen as both a version management implementation (an earlier version, "*Practical Ethics,*" was published in *V&A Conservation Journal* 56 (2008): 14–15) and as an echo of the content revision management tool employed by the wiki on which the text has been written.

3. The title of a 1964 short story by J.G. Ballard where a man goes into mental and physical decline whilst hiding in the decaying buildings of an island once used for testing nuclear weapons. This, and all other section headings, are deployed as a playful exchange made within the text, often to indicate where the author sees that a form of cultural production is rooted in particular technological developments.
4. It is well noted elsewhere how 'scientific analysis' can be successfully applied when concerned with isolated phenomena, but less able to respond when facing complexity and revealing its probabilistic nature in its specifications when matched with real world behaviours. See S. Muñoz Viñas, *Contemporary Theory of Conservation* (Oxford: Elsevier, 2005) 121–129.
5. An essentialist epistemology suggests that how knowledge is characterized represents a key driving force of a knowledge economy. Thus, for example, the influence of the heterogeneous backgrounds of all conservators as people is insignificant when compared to their socialization into the prevalent knowledge characteristics of their discipline.
6. 'Flip' and the following 'Flop,' refer to a flip-flop device, a device capable of either one of two stable states, but not the two together, and one which underpins all computational technology. Flip-flops along the Terminal Beach.
7. F. Matero, "Ethics and Policy in Conservation", *Getty Conservation Institute Newsletter*, Volume 15, Number 1 Spring, (2000) 6.
8. See: International Council of Museums *Code of Ethics for Museums*, 2006 '2.24 Collection Conservation and Restoration: The museum should carefully monitor the condition of collections to determine when an object or specimen may require conservation-restoration work and the services of a qualified conservator-restorer. The principal goal should be the stabilization of the object or specimen. All conservation procedures should be documented and as reversible as possible, and *all alterations should be clearly distinguishable from the original object or specimen*' (Author's italics).
9. Christopher Caple's RIP Model (revelation, investigation and preservation) presents a similar triadic graph for mapping specific treatments: C. Caple, *Conservation Skills: Judgement, Method and Decision Making* (London: Routledge, 2000) 34.
10. The group name of a late 1960s Turin-based Italian autonomist movement, and its eponymous newspaper, that sought to establish a way of living within a wider community but governed by its own system of rules.
11. ICOM 2006 2.24.
12. Re-entering the discussion on authenticity with a renewed characterization, one that implicitly accepts the malaise of Rationalism, by which it is claimed, in another J.G. Ballard short story of the same title, scientific endeavour is rendered as futile.
13. 'Digital preservation combines policies, strategies and actions to ensure access to reformatted and born digital content regardless of the challenges of media failure and technological change. The goal of digital preservation is the accurate rendering of authenticated content over time.' Long Definition in "Definitions of Digital Preservation," prepared by the Preservation and Reformatting Section, Working Group on Defining Digital Preservation, ALA Annual Conference, Washington, DC, June 24, 2007 <http://www.ala.org/ala/alcts/newslinks/digipres/index.cfm> Accessed April 2008.

14. Time-based art is not alone in its defiance, for example, Navajo sand paintings and Amazonian baskets equally challenge the means of authenticity's description. See C. Classen and D. Howes, "The Museum as Sensescape: Western Sensibilities and Indigenous Artefacts", *Sensible Objects: Colonialism, Museums and Material Culture*, eds. E. Edwards, C. Gosden and R.B. Phillips) (Oxford and New York: Berg Publishers Ltd, 2006).
15. P. Laurenson, "Authenticity, Change and Loss in the Conservation of Time-Based Media Installations", *Tate Papers* (London: Tate, 2006) (http://www.tate.org.uk/research/tateresearch/tatepapers/06autumn/laurenson.htm). Accessed April 2008.
16. A play on the word 'plot' with reference to how carefully any narrative argument sets itself out. Every scene, every line works but then later you see how you were set up: nobody says anything that doesn't become important to the plot later.
17. A 1979 UK road movie by Chris Petit, widely seen as a film that successfully diagnoses its moment, and where each film cut also cuts the soundtrack.
18. See: A. Zander Rudenstine and T.P. Whalen, "Conservation Documentation in Digital Form: A Dialogue about the Issues", *Getty Conservation Institute Newsletter* 21.2 Summer (2006). <http://www.getty.edu/conservation/publications/newsletters/21_2/news_in_cons.html> Accessed April 2008. For the full terms and archive of the documented discussions from New York (2006) and London (2007), see: (http://mac.mellon.org/issues-in-conservation-documentation). Accessed April 2008.
19. Rudenstine and Whalen 27 (http://www.getty.edu/conservation/publications/newsletters/21_2/news_in_cons.html). Accessed April 2008.
20. The eponymous theory that there is a very large, perhaps infinite, number of universes and that everything that could possibly happen in our universe (but doesn't) does happen in some other universe(s).
21. This paragraph is paraphrased from passages found in the Issues in Conservation Documentation: Digital Formats, Institutional Priorities, and Public Access – Report of a meeting held under the auspices of the Andrew W. Mellon Foundation at the Metropolitann Museum of Art, New York, April 27, 2006 (http://mac.mellon.org/issues-in-conservation-documentation/issues-in-conservation-documentation/2006_new_york_dossier.pdf). Accessed April 2008.
22. *Back to the Future* is a 1985 science fiction film directed by Robert Zemeckis, precisely executed so that it vindicates its own formulaic construction.
23. See: *Issues in Conservation Documentation: Digital Formats, Institutional Priorities, and Public Access, London* – Report of a meeting held under the auspices of the Andrew W. Mellon Foundation at the British Museum, May 25, 2007 (http://mac.mellon.org/issues-in-conservation-documentation/2007_london_dossier.pdf). Accessed April 2008.
24. In the first Shrek movie of 2001, Shrek tries to explain to his friend Donkey that actually ogres are like onions, in that they both have layers beneath the surface.
25. Various early UK ventures in the open publishing of state-funded information include BBC/Channel 4's Creative Archive, an open national television archive project (http://creativearchive.bbc.co.uk/ and the open national geo-data project: http://okfn.org/geo/).

7

Conservation Principles in the International Context

Jukka Jokilehto

The purpose of this chapter is to examine the development of conservation principles with particular emphasis on authenticity and integrity and the change of emphasis from material to intangible aspects of heritage. Examining the general trends in the development of the principles from the statements by William Morris to the international charters of the twentieth century, one can observe a constant concern for the preservation of the original material evidence in heritage resources. On the other hand, by the end of the twentieth century, the extension of the notion of heritage has come to include the entire living environment with its cultural traditions and changing life styles. As a result, the concept of heritage conservation is thus becoming less static in reference to historic material, and rather more dynamic with reference to culturally sustainable management of heritage resources, taking into account their tangible and intangible dimensions. Conservation principles should not be seen in isolation, but rather as highlights of the application of the theory of restoration-conservation. At the same time, a return to the internationally adopted conservation principles, and the examination of their validity in the different cultural contexts, keeps alive an interest in, and the appreciation of, what is considered as heritage in each case and what should be conserved.

The modern concept of conservation has appeared gradually with the modern *Weltanschauung* (world view). In this sense, it is closely related to the so-called western world, the world that developed from the Mediterranean region in the fifteenth century. A crucial period in the foundation of these concepts was the Age of Reason, the eighteenth century. In the history of philosophy, this is the period of Immanuel Kant, whose contribution has deeply penetrated modern thought. There were also other thinkers such as Giovanni Battista Vico, a Neapolitan lawyer who associated truth to people's cultural history, and Johann Gottfried Herder, a German philosopher and writer who further contributed to the concept of cultural pluralism. The period forms the setting from where German Romanticism and the French Revolution emerged at the end of the century, both contributing in different ways to the modern world view and to today's conservation culture. The early debates on restoration vs. conservation that took place in Western and Southern Europe in the eighteenth and nineteenth centuries were instrumental for the further definition of some of the basic principles. Here we can hear the voices of Victor Hugo and Adolphe Napoleon Didron in France, John Ruskin and William Morris in England, and Alois Riegl in the Austro-Hungarian empire, just to mention a few who influenced this development.

On March 5, 1877, a letter by William Morris was published in *The Athenaeum*, where he protested against the destruction of the Minster of Tewksbury due to restoration by Sir Gilbert Scott, proposing that 'an association should be set on foot to keep a watch of old monuments, to protect against all "restoration" that means more than keeping out wind and weather, and, by all means, literary and other, to awaken a feeling that our ancient buildings are not mere ecclesiastical toys, but sacred monuments of the nation's growth and hope.'[1] Morris became secretary to the Society for the Protection of Ancient Buildings (SPAB) that was then established and whose principles of conservation of ancient buildings were to influence attitudes not only in England and Great Britain, but in many other countries as well. In the *Manifesto* of the Society, echoing the words of Ruskin, Morris wrote:

> If, for the rest, it be asked of us to specify what kind of amount of art, style, or other interest in a building, makes it worth protecting, we answer, anything which can be looked on as artistic, picturesque, historical, antique, or substantial: any work in short, over which educated, artistic people would think it worthwhile to argue at all. It is for all these buildings, therefore, of all times and styles, that we plead, and call upon those who have to deal with them to put Protection in the place of Restoration, to stave off decay by daily care, to prop a perilous wall or mend a leaky roof by such means as are obviously meant for support or covering, and show no pretence of other art, and otherwise to resist all tampering with either the fabric or ornament of the building as it stands; if it has become inconvenient for its present use, to raise another building rather than alter or enlarge the old one; *in fine* to treat our ancient buildings as monuments of bygone art, created by bygone manners, that modern art cannot meddle with without destroying. Thus, and thus only, shall we escape the reproach of our learning being turned into a snare to us; thus, and thus only can we protect our ancient buildings, and hand them down instructive and venerable to those that come after us.[2]

The idea of protecting historic buildings as cultural heritage was not only an issue of specific countries, but it emerged rapidly as a worldwide conservation movement. One of the early international conferences on this issue was organized in Athens in 1931 by the International Office of Museums. This was a non-governmental organization created after the First World War by the International Office of Intellectual Cooperation (later re-established as UNESCO) in the framework of the League of Nations (later United Nations). The *General Conclusions of the Athens Conference* expressed some of the fundamental principles that had matured at the time:

> Whatever may be the variety of concrete cases, each of which are open to a different solution, the Conference noted that there predominates in the different countries represented a general tendency to abandon restorations *in toto* and to avoid the attendant dangers by initiating a system of regular and permanent maintenance calculated to ensure the preservation of the buildings.
>
> When, as the result of decay or destruction, restoration appears to be indispensable, it recommends that the historic and artistic work of the past should be respected, without excluding the style of any given period.
>
> The Conference recommends that the occupation of buildings, which ensures the continuity of their life, should be maintained but that they should be used for a purpose which respects their historic or artistic character.[3]

After the Athens conference, the development of conservation ideas continued. This was particularly interesting in Italy, where Gustavo Giovannoni, one of the participants in Athens and a chief architect at the time of Mussolini, wrote an Italian charter of restoration, *Norme per il restauro dei monumenti* (1931–1932). Here he echoed the ideas of Ruskin and Morris as well as the Athens conclusions:

> . . . considering that restoration work should take into account but not eclipse even partially various types of criteria: that is to say the historic reasons whereby none of the phases which comprise the monument should be eliminated or falsified by additions which might mislead scholars, nor should the material brought to light through analytical research be lost; the architectural concept which aims at the correct rehabilitation of the monument and, whenever possible, to a unity of form (not to be confounded with a unity of style); the criteria based on public sentiment, on civic pride, on its memories and nostalgia; and finally on what is considered essential by the appropriate administration in line with the means available and eventual practical use.[4]

Between the two World Wars, there was a further advance of the principles and practice of restoration and conservation. While the foundations were already laid in the 1930s, the policies had a further important development as a result of the devastating experience of the Second World War. In Italy, the contribution of Giulio Carlo Argan and Cesare Brandi, two art historians, was fundamental to the development of restoration as an autonomous field based on the recognition and critical assessment of the significance and values of the historic-artistic works. In his *Theory of Restoration* (1963), Brandi focused on the restoration of the work of art, and emphasized this as a critical process. He defined restoration as: 'the methodological moment in which the work of art is recognized, in its physical being, and in its dual aesthetic and historical nature, in view of its transmission to the future.'[5] Regarding the work of art, he stressed that the aim of restoration should be: 'to re-establish the potential oneness of the work of art, as long as this is possible without committing artistic or historical forgery, and without erasing every trace of the work of art's passage through time.'[6] The 1950s and 1960s were a period of post-war reconstruction, where the treatment of damaged or destroyed historic buildings became one of the important issues to decide. The principles that had emerged in the early part of the century did not seem sufficient. In fact, the theory of Cesare Brandi, the first director of the newly founded Italian Institute of Restoration in Rome, found its full justification in the necessity to safeguard and restore severely damaged works of art.

In May 1964, in Venice, on the invitation of the Italian government, UNESCO sponsored the Second International Congress of Architects and Technicians of Historic Monuments. Here the experts drafted the *International Charter for the Conservation and Restoration of Monuments and Sites, the Venice Charter*, which summarized some of the fundamental principles of restoration and conservation. The charter stressed that the question was of living witnesses of age-old traditions, and that these formed a common heritage of humankind based on shared values. At the same time, the charter reminded us that it was our duty to hand on these monuments in the full richness of their authenticity. Much of the thinking behind the charter came from Brandi's *Theory of Restoration,* which had just been published the previous year. What started becoming clear was also that the principles should not be interpreted in isolation but should rather be taken as highlights of a new culture based on the recognition and safeguarding of historic objects, monuments and sites. This culture was seen as the product of the modern world society, which went beyond the local cultural specificities. Thus restoration theory provided a methodology for the recognition and safeguarding of heritage and was considered to have universal validity. The emerging principles were to be understood as integral parts of the theory and as an expression of the modern conservation-restoration culture.

The fundamental ideas of the Venice Charter could be synthesized in terms of the notions of authenticity and integrity, even though these concepts may not have been so clearly defined. *Restoration* was seen as a specialized operation, which required specific training. It meant also that the profession of a restorer was becoming recognized as a profession and not only as an additional task amongst others. It should therefore not be a surprise that from this time on an increasing number of training opportunities was offered in various countries, particularly in Europe but also, for example, in Turkey and USA. One of the important references for the development of specialized training programmes in the different fields of expertise was the International Centre for the Study of the Restoration and Preservation of

Cultural Property, ICCROM, which was founded by UNESCO in 1956 and based in Rome. Here it had close collaboration with the Italian Institute of Restoration, where Brandi was Director. Professor Paul Philippot, then Assistant Director of ICCROM, was one of the key authors of the Venice Charter.

The Venice congress also recommended the establishment of ICOMOS, the International Council on Monuments and Sites. This took place in Poland the following year, and the new organization decided to take the Venice Charter as its founding ethical commitment. One of its principal tasks was to work for the clarification of the principles and diffuse examples of good practice to conservation professionals in the field of the built heritage. Through its international conferences and symposia it has significantly contributed to raising awareness and building attitudes. A parallel task in the field of collections and museums was taken by ICOM, the International Council of Museums, which emerged from the re-foundation of the International Museums Office after the Second World War. The international charters adopted by ICOMOS include the *Florence Charter on Historic Gardens* (1981), the *Charter for the Conservation of Historic Towns and Urban Areas* (1987), the *Charter for the Protection and Management of the Archaeological Heritage* (1990), the *Charter on the Protection and Management of Underwater Cultural Heritage* (1996), the *International Tourism Charter* (1999), and the *Charter on the Built Vernacular Heritage* (1999). In addition, there are documents called principles, including: the *Principles for the Preservation of Historic Timber Structures* (1999), the *Principles for the Preservation and Conservation/Restoration of Wall Paintings* (2003), and the *Principles for the Analysis, Conservation and Structural Restoration of Architectural Heritage* (2003).

ICCROM as an interdisciplinary inter-governmental organization has contributed to the development of training programmes in all fields, while UNESCO, through its international recommendations and conventions, has provided the general framework. In 1972, it adopted the *Convention concerning the Protection of the World Cultural and Natural Heritage, the World Heritage Convention*, which has since become one of the most powerful instruments in promoting the safeguarding and management of heritage resources worldwide, with respect to the conservation principles expressed in the international doctrine. The definition of the cultural heritage in the convention was limited to monuments, groups of buildings and sites. However, the field has since been broadened, including particularly cultural landscapes and cultural routes. While in the early years of the World Heritage List, emphasis was given to monumental heritage, the trend has since been to more vernacular types of properties. Also, rather than nominating single buildings, the tendency has been towards larger areas, historic towns and landscapes. In this context the notions of authenticity and integrity have been newly emphasized becoming popular topics for research and conference papers.

From the nineteenth century, the main emphasis in conservation principles had generally been given to the physical remains of ancient sites or material objects. Over the second half of the twentieth century, the trend has however been towards the immaterial or intangible aspects of human creativity. Such issues were already included under legal protection in Japan and Korea relatively early. An international recognition of these trends has come from UNESCO with the *Proclamation of Masterpieces of the Oral and Intangible Heritage of Humanity* (1998), the *Universal Declaration on Cultural Diversity* (2001), and the *Convention for the Safeguarding of the Intangible Cultural Heritage* (2003). Not only does this broadening of the definition of cultural heritage involve issues that earlier were given little or no attention, but it also has given rise to new challenges to the definition of conservation principles.

Questions are now raised about the universality of internationally adopted conservation principles as well as about the significance of authenticity and the justification of emphasis on the preservation of historical material over reconstruction. For example, in South Korea, Professor Seung-Jin Chung has criticized the Venice Charter as being too European in its emphasis and too much oriented on visual beauty through its material substance. In his view, the East-Asian societies are determined more in relation to the spiritual and naturalistic qualities of their culture and architecture.[7]

In China, the conservators have taken the initiative to prepare a set of national guidelines that integrate conservation and management, *Principles for Conservation of Heritage Sites in China*, published in Chinese in 2000 and in English in 2002. The project was based on collaboration between China's State Administration for Cultural Heritage, the Getty Conservation Institute (USA), and the Australian Heritage Commission. These principles mainly refer to ancient monuments and historic buildings, and they are based on the existing guidelines or charters, such as the *Venice Charter* (1964) and the *Burra Charter* by the Australian National Committee of ICOMOS (rev. 1999). Much attention is given to the process of assessment of a property and its management. In China, however, not everybody agrees with the international principles. As a result, in 2005, a group of technicians has prepared the so-called *Qufu Declaration* resulting from a symposium on traditional architecture that discussed the continuation of traditional crafts and skills. The idea here is that when there are the conditions and sufficient evidence, it should be acceptable and even encouraged to rebuild a damaged historic structure in its most complete form.

In Japan, as well, the debate has started stressing the specific characteristics of Japanese architecture and settlements, and their relationship with their context. An example of this is the so-called *Machinami Charter*, adopted by Japanese ICOMOS in 2000. The concept of *machinami* is usually translated as historic town, but the word has a much more specific and subtle meaning relating to the tangible and intangible factors, the physical and spiritual aspects, that define a settlement, its buildings, its activities, and its natural environment. Great interest has also been shown in traditional animistic religious practices, exploring the principles that should guide their recognition, and eventual safeguarding and regeneration. Here the issue, in fact, is that aspect of intangible heritage involving living traditions and spiritual practices, which are still associated with specific places and their surroundings.

In 2004, the Indian National Trust for Art and Cultural Heritage (INTACH) has adopted the *Charter for the Conservation of Unprotected Architectural Heritage and Sites in India*. The innovative aspect of this charter is that, rather than focusing on protected cultural heritage, it is concerned with non-protected cultural heritage, i.e. historic buildings and urban areas that contain traditional buildings and structures, which are subject to transformation and risk of being lost due to development. The purpose in conserving such built heritage is to keep the sense of place and the character of the environment. It offers the opportunity not only to conserve the past, but also to define the future. It provides alternate avenues for employment and a parallel market for local building materials and technologies, which needs to be taken into account when resources for development are severely constrained. This living heritage also has symbiotic relationships with the natural environments within which it originally evolved. Understanding the interdependent ecological network and conserving it can significantly contribute to improving the quality of the built environment as well.

The notion of **being authentic** could be taken as meaning being truthful. In the context of the World Heritage Convention authenticity has been given a high profile, resulting in the organization of an expert conference in Nara, Japan, in 1994. This reflection has since continued, and there are specific attributes that are often listed as references for the definition of authenticity, such as design, material, workmanship and setting, but also the spirit of the place, its use and other cultural and social parameters. At the same time, it should be clear that authenticity should not be measured in reference to one sole parameter, but based on a critical judgement of the whole. The *Nara Document on Authenticity* of 1994 highlights cultural pluralism and the respect of diversity of values as a spiritual and intellectual heritage, as well as recognizing the 'intangible heritage,' already present in the Japanese Law for the Protection of Cultural Properties of 1954. Generally speaking, authenticity can be seen in relation to different parameters, and particularly to the following:

a. *Qualitative judgement* in reference to a work considered as a result of human creative process and therefore autonomous and not a copy;
b. *Legal verification* in reference to the material truth of an object as an historic document;
c. *Social-cultural traditions* and inculcation of value judgements.

The modern historical consciousness was not born by accident, but was the result of a process. The evolution of the modern concept of authenticity should be seen in such a perspective. Our limitation in judging the differences of other epochs lies in our own *Kunstwollen*, which is based on our present-day values. Modern thinkers, including Martin Heidegger and Cesare Brandi, have proposed that the work of art is the result of a human creative process. In consequence, the significance of a work of art is in its intangible or immaterial dimension, of which the material becomes a carrier. In order to understand and fully appreciate a work of art, there is a need for a process of recognition of its significance. Through such a process, the observer will thus re-appropriate the meaning of a work of art in his/her mind. It is through this process of becoming conscious that we also prolong the life of a work of art, both in its conception and its physical aspects. The truth of creativity or the notion of authenticity in relation to the truthfulness of the creative process should be the basis for the qualitative judgement of authenticity.

Referring to the concept of authenticity, in this sense, it seems useful to note the definition by Dr Paul Philippot (Director Emeritus of ICCROM): 'the authenticity of a work of art is the internal unity of the mental process and of the material realization of the work.'[8] Such a mental process should not necessarily be seen in reference to an 'autograph,' understood in its modern conception. Many works, today conceived as 'works of art,' are the result of the work of many people. This could either be a teamwork creating a *Gesamtkunstwerk* at a particular time or the contribution by different hands over time. Therefore, especially in the second case, such a work may only have been possible through the involvement of several generations. It is mainly since the period of Romanticism that specific importance has been given to an 'autograph' by an individual artist. Before this time, traditional arts and crafts were generally based on a creative imitation of what existed. This was seen not only as the proper way of doing things but also useful. One could prepare a replica for one's own benefit as well as for the interest of others. Imitation guaranteed the continuity of traditions, as well as being a form of safeguarding the skills and the know-how of the previous generations. Copying was also used as a way to learn the techniques and the aesthetic 'canons' that had been refined by previous masters. Traditional approaches based on models may or may not stimulate the individual's imagination, but this is unlikely to be the intention behind them. In contrast, the modern idea of creativity is based on individuality of the imagination of the artist.

Authenticity is part of the culture of our time, and it is often contrasted with the concept of false or fake. This association might need to be placed in perspective, and reconsidered especially regarding the idea of a copy. The Ise Shrine in Japan, which is reconstructed periodically, is often taken as an example of traditional 'Japanese restoration.' This Shinto sanctuary actually consists of two sites with dozens of shrines – some of them rather large. Several of these are part of the ceremonial reconstruction process occurring circa every 20 years. It is not only the buildings, but even all the equipment, such as the textiles, and the arms that are periodically renovated. This operation, however, is not considered 'restoration' by the Japanese, but rather a traditional, religious ritual. As in many other cases in Oceania, the significance is more in the ritual performance than in the material that is often lost as part of the process. The significance of the process of rebuilding the Ise Shrine is in the traditional continuity as a response to genuine religious feelings. That is also where its authenticity lies.

The *Nara Document on Authenticity* declares: 'Conservation of cultural heritage in all its forms and historical periods is rooted in the values attributed to the heritage. Our ability to understand these values depends, in part, on the degree to which information sources about these values may be understood as credible or truthful.'[9] It is here that we can look for authenticity in a particular historic object or structure. On the other hand, taking the contrary, i.e. something inauthentic or fake, this can be defined as an artefact where the sources of information have been altered or falsified. A fake can be intentional, for example, for marketing purposes, or it can be non-intentional. The latter could easily be applied to archaeological sites, where reconstructions may cover the original evidence, obstructing a critical study of the sources, or where the restoration reflects interpretations based only on partial information and understanding of the site. The same could happen in historic urban areas, where restoration and rebuilding may have become a fashion, often justified by tourism, and where the urban fabric, in reality, is falsified through such projects. One of the problems of our time is the enormous quantity of secondary information, which may easily deviate our attention from the original. It is generally accepted that verifying truth in an historic monument is only possible on the original – not on a modern replica.

Another concept that has emerged within the World Heritage context is the **condition of integrity**. First required only for natural heritage, the World Heritage Committee has decided that it should be verified also in relation to cultural heritage, as indicated in the 2005 edition of the *Operational Guidelines for the Implementation of the World Heritage Convention.* Here this notion is defined as 'a measure of the wholeness and intactness of the natural and/or cultural heritage and its attributes.'[10] In order to fulfil the condition of integrity, a property should thus include all elements necessary to express its significance and be of adequate size to ensure the complete representation of the features and processes which convey its significance. Finally, the property should not suffer from adverse effects of development and/or neglect. The question of integrity can become intricate in the case of complex historic properties. In fact, the Venice Charter stresses the importance of historical integrity, stating that 'the valid contributions of all periods to the building of a monument must be respected, since unity of style is not the aim of a restoration.'[11] On the other hand, the charter also emphasizes the need to find a reasonable balance in the aesthetic integrity stating that 'replacements of missing parts must integrate harmoniously with the whole, but at the same time must be distinguishable from the original so that restoration does not falsify the artistic or historic evidence.'[12] Furthermore, it is noted that, while the charter does not go into any depth regarding the conservation of historic areas, it nevertheless underlines that 'The sites of monuments must be the object of special care in order to safeguard their integrity and ensure that they are cleared and presented in a seemly manner.'[13]

The globalizing world of the twenty-first century offers new challenges to the conservation of heritage. The notion of 'cultural heritage' has been broadened resulting from the recognition of a great diversity of resources as heritage, including cultural landscapes or just places of memory. Heritage is qualified in its diversity and in its material and immaterial (tangible and intangible) aspects. Due to the present-day holistic approach and the need to recognize the specificity of each place, conservation theory must necessarily be seen as a methodology based on critical judgement, and generally integrated with the planning and management processes. From a static action restoring a place in a particular form, conservation is increasingly seen as a dynamic process. It should take into account the evolving values and involve the different stakeholders. Within such a context, the question can be raised whether it is possible to establish some universal principles at all! The newly revived debate worldwide, including Africa, India, China and Japan, would seem to confirm that there are such principles. At the same time, restoration cannot be based on standard solutions. Rather, each intervention needs to take into account the cultural specificity of the place. It is therefore necessary that internationally adopted principles be integrated with appropriate guidelines taking into account the cultural and historical context of each place, a process that is already giving its first results.

References for conservation guidelines

N. Agnew and M. Demas, eds., *Principles for the Conservation of Heritage Sites in China* (Los Angeles: The Getty Conservation Institute, 2002).

ICOMOS, *International Charters* for Conservation and Restoration, Monuments and Sites I, ICOMOS, Munich 2004. Available from: <http://www.icomos.org/>.

J. Jokilehto, 1996. "International standards, principles and charters of conservation". S. Marks, ed. *Concerning Buildings, Studies in Honour of Sir Bernard Feilden*. (Oxford: Architectural Press) 55–81.

J. Jokilehto, *A History of Architectural Conservation*. (Oxford: Butterworth-Heinemann, 1999)

UNESCO Conventions and Recommendations. Available from: <http://portal.unesco.org/>.

UNESCO, Basic Texts of the 1972 World Heritage Convention. (Paris: UNESCO World Heritage Centre, 2005)

Notes

1. Norman Kelvin, ed., *The Collected Letters of William Morris*, Vol. I, 1848–1880 (Princeton, New Jersey: Princeton University Press, 1984) 351f.
2. Kelvin, 359f.

3. General Conclusions of the Athens Conference, Athens, 1931, 26 September 2008 (http://www.icomos.org/athens_charter.html).
4. "Carta del restauro italiana (Roma 1931)," Ministero per i Beni Culturali e Ambientali, Ufficio Studi, La conservazione dei beni culturali nei documenti italiani e internazionali, 1931–1991, Guglielmo Monti, ed. (Rome: Istituto Poligrafico e Zecca dello Stato, 1995) 12–13.
5. Cesare Brandi, *Theory of Restoration* (Rome: Nardini Editore, 2005) 48. 'Il restauro costituisce il momento metodologico del riconoscimento dell'opera d'arte, nella sua consistenza fisica e nella sua duplice polarità estetica e storica, in vista della sua trasmissione al futuro.' In Cesare Brandi, *Teoria del restauro* (Rome: Editori Riuniti, 1963) 34.
6. Brandi, *Theory* 50. 'Il restauro deve mirare al ristabilimento della unità potenziale dell'opera d'arte, purchè ciò sia possibile senza commettere un falso artistico o un falso storico, e senza cancellare ogni traccia del passaggio dell'opera d'arte nel tempo.' In Brandi, *Teoria*, p. 36.
7. Seun-Jin Chung, "East Asian Values in Historic Conservation," *Journal of Architectural Conservation*, Volume 11, Number 1 (2005): 55–70.
8. Paul Philippot, Personal interview, January 1995. See: J. Jokilehto, "Viewpoints: The debate about authenticity," *ICCROM Newsletter* Volume 21 (July 1995): 6–8.
9. *Nara Document on Authenticity*, 1994, art. 9.
10. UNESCO, *Operational Guidelines for the Implementation of the World Heritage Convention* (Paris: UNESCO, 2005) par. 88.
11. Venice Charter, 1964, article 11.
12. Venice Charter, 1964, article 12.
13. Venice Charter, 1964, article 14.

8

The Concept of Authenticity Expressed in the Treatment of Wall Paintings in Denmark

Isabelle Brajer

Introduction

The concept of authenticity is closely linked to the debate about contemporary conservation/restoration principles in the Western cultural tradition because our profession is currently experiencing the cultivation of two parallel trends regarding the main focus and meaning of our work. Modern conservators have primarily coupled authenticity with the physical substance of objects. This attitude can be seen in the ongoing and longstanding concentration on the preservation of the material fabric and emphasis on preventive treatment. At the same time, new ideas about the *raison d'être* of our profession postulate that the ultimate goal of conservation is not to preserve the material aspects of a particular object, but to retain or improve the meaning it has for people.[1] On the one hand, numerous publications underscore scientifically founded preservation as the supreme principle of our profession today. On the other, a clear recent trend embraces community involvement and communication in conservation and restoration projects, where the focus is often shifted to aesthetic issues involving presentation and appearance, or symbolic values inherent in the object. The question is whether the two currents that coexist today have emerged because of a shifting understanding of authenticity, or whether the trends have had an influence on the broadening of our comprehension of authenticity. In the given circumstances the two trends can be considered both complimentary and contentious. For example, advocates of minimal treatment can argue that more effort should be placed on communicating our goals to the general public in order to improve its understanding of our work. Alternatively, situations will occur where the improvement of public understanding of the object necessitates a certain degree of aesthetic processing, which can compromise its material authenticity.

The field of conservation is affected by transitions in the general concept of culture: where the emphasis is shifted from the product to the process; where intangible values are treated on equal footing with tangible values; where popular interests are superseding elitist tendencies. Just as the concept of what culture is has undergone a transition in time, becoming more inclusive, so has the concept of authenticity. So, given its diffused and multi-faceted character, what is authenticity? Jukka Jokilehto defines authenticity as a condition inherent in an object,[2] a measure of 'truthfulness of the internal unity of the creative process and the physical realization of the work, and the effects of its passage through historical time.'[3] The affirmation of authenticity in the object is a basis for the measurement of relevant cultural values (age value, artistic value, memorial value, historical value, cult value, exhibition value). On the other hand, the values themselves play an important role in the attribution of the status of authenticity.[4] As a result, there is a contradictory situation in some objects or monuments, including ones entered on the World Heritage List (such as the historic centre of Warsaw in Poland, which was reconstructed after total destruction during WWII) that are considered 'authentic' as a result of their reconstructions,[5] an illustration of shifting the emphasis from preserving genuine material substance to the thoughts and emotions that the site or general cultural landscape evokes.[6]

The relatively recent focus on authenticity over the past decade or so (largely, as a result of the Nara Conference on Authenticity in 1994)[7] has highlighted the complex sides of this issue, entailing discussions of various values and functions, including the more intangible properties of a work of art, such as its emotional qualities. This abundance is expressed in such concepts as: authenticity of form and design, materials and substance, use and function, traditions and techniques, location and setting, spirit and feeling, and other factors, such as appearance, context and intention.

Examining the issue of authenticity and how it relates to conservation – restoration practice is an enormous and difficult task. Perceptions of authenticity are relative. In fact, each culture and entity accords authenticity a different meaning – a meaning that undergoes transitions over time.[8] Moreover, the artistic meaning and historical function of objects change, and also the way we perceive them[9] which affects the interpretation of authenticity. There have been many interesting and excellent publications on the subject of authenticity, the global scope of which this contribution does not aspire to. This chapter hopes to fill a niche in the authenticity discourse by focusing on wall paintings, which can be described as cutting a thin slice from the multi-layered cake of authenticity. Furthermore, the deliberations will concentrate on the aesthetic treatment of medieval wall paintings in Danish churches, which is like taking one forkful of that slice. However, the intention of narrowing the focus to this specific area can be supported by the fact that it is often through the interface of aesthetics (appearance and presentation) that broader concepts of authenticity can be illustrated. The information pertaining to Danish wall paintings will undoubtedly not be applicable in its entirety to other objects of cultural heritage, even to wall paintings in other European countries – which itself is a demonstration of how complex the concept of authenticity is. Issues of respect for the artist's intent (as problematic as that may be) and the often accompanying connoisseurial interpretations have little impact in cases where wall paintings, executed mostly by anonymous painters or craftsmen, have undergone profound physical changes primarily due to the fact that they were over-painted with limewash, and then uncovered again decades later, as is the case in Denmark. This illustrates how local (regional) conditions and circumstances affect the authenticity debate, even within one specific area of cultural expression.

Authenticity in the nineteenth century

The modern concept of authenticity in Europe has its roots in the new historical awareness that formed in the eighteenth century, and is an offspring of Romanticism and its focus on national identity and traditions.[10] As in many other countries, Denmark experienced a Romantic medieval revival. Interest in the wall paintings in medieval churches was a result of these growing nationalistic feelings, particularly after the loss of its southern territory to Germany in 1864, and can be linked even earlier to the publication of such literature as B.S. Ingemann's *Valdemar the Great and his Men* (1824), which described Denmark's days of glory and power when it emerged as a great nation in the twelfth century.[11] The numerous wall paintings from the medieval period, which were rediscovered in the second half of the nineteenth century after existing for decades under a thick layer of limewash, testified to this period of magnificence.

These decorations brought the past to life. However, many survived in a fragmentary or poor condition due to changes in the architectural structure, and also to the rather harsh treatment (over-painting with limewash and subsequent uncovering) to which they had been subjected. In the process of retrieving this past, nineteenth-century restorers strove, as far as they could, to complete the paintings, recreating an idealized original state. There are numerous examples of re-painting and speculative reconstructions, the extremes of which can be illustrated by the restoration that took place in 1888 in Engum Church, where fragments of a Late-Romanesque painting were uncovered on the soffit of the chancel arch. The decoration originally portrayed Cain and Abel on the flanks, but only the upper part of Abel survived, and the entire figure of Cain was gone (Figure 8.1a and b). Not only did the restorer reconstruct the bottom half of Abel, but he also recreated Cain by copying the figure from a similar composition from another church (Tyvelse). Restorers in the second half of the nineteenth century, working in the artist-restorer tradition, focused on presenting truthfulness in terms of the general pictorial content or composition in order to present a more complete picture. It was of less importance that details, such as drapery folds, positions of arms, and even entire figures, could not be reproduced with absolute validity, a view that clearly recalls Eugène-Emmanuel Viollet-le-Duc's attitude. According to Viollet-le-Duc: 'To restore an edifice means neither to maintain it, nor to repair it, nor to rebuild it; it means to re-establish it in a finished state, which may in fact never have actually existed at any given time.'[12] In the case of Engum, it was considered legitimate to replace the missing figure of Cain opposite Abel since it was believed to have been there before the painting was damaged, especially when there existed a model to copy from. The restorer, Jacob Kornerup – active in the years 1862–1904 – received the blessing of the antiquarian authorities to perform this reconstruction. The artist-restorers strove to imitate the original painting style in their reconstructions. But more often than not their interpretations did not relay the same artistic qualities as the original. In fact, the style of the concurrent art movement, with its focus on scrupulous execution, often imposes a nineteenth-century aesthetic on many medieval wall paintings restored in this period.

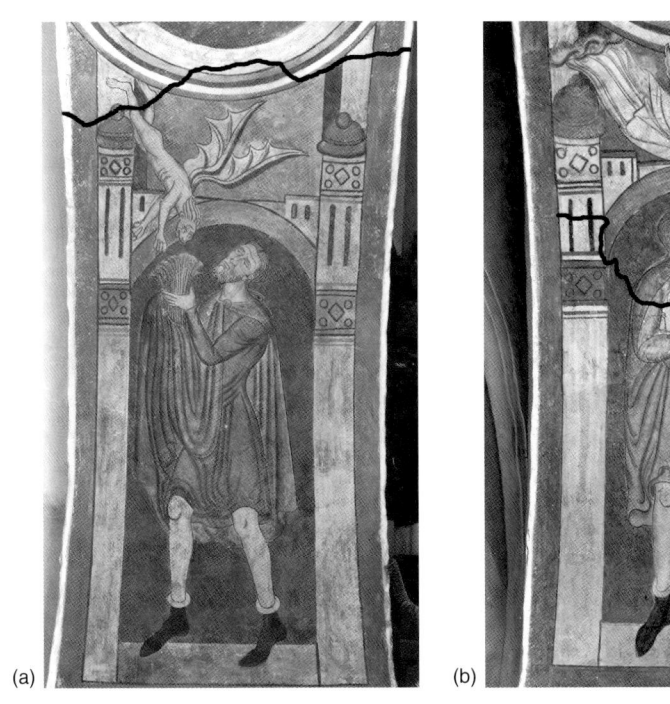

Figure 8.1a and b *Wall paintings depicting Cain and Abel in Engum Church. When the paintings were found in 1888, only the fragments above the black line were intact. The figures were reconstructed using a similar composition in another church as a template. Photo reproduced with permission from The National Museum of Denmark.*

Early twentieth-century restoration

The artist-restorer tradition, far from being unique to Denmark, trickled on well into the first half of the twentieth century, somewhat longer than in some European countries. Eigil Rothe (active 1897–1929) and Egmont Lind (active 1926–1966) were very skilled imitators of medieval brushwork. Many of their reconstructions can only be differentiated from the original parts of the paintings on close scrutiny and can easily confuse even a professional.

Eigil Rothe was the first Danish restorer to distinguish between unethical aesthetic processing and ethical treatment. The former, which he called 'restorations,' would typically involve unsubstantiated reconstructions or over-painting that smacked of nineteenth-century aesthetics, in contrast to the latter, which he called 'preservations.' However, the contemporary meaning of these terms does not reflect Rothe's thinking. Rothe's concept of 'preservation' referred to the original painter's expression and style, but not necessarily to the original substance. For Rothe, preserving the authenticity of a painting demanded a highly skilful and disciplined approach, in order to achieve a faithful rendering of the artist's original brushstrokes within the lacunae. Rothe sometimes reconstructed entire faces, using other undamaged figures in the same painting as models – this was not considered by him to be speculative reconstruction (larger lacunae, where there was no indication of the original contents, were toned down with a tinted limewash). He also supported the idea of extensive re-painting of the original areas if the colours were faded and weak. However, Rothe felt it was wrong to reconstruct and re-paint in a way that would imitate the pristine condition of the paintings, as they appeared when they were created. Indeed, he believed that the reconstructed areas should imitate the muted appearance of a painting that had undergone the process of being covered with limewash, and subsequently uncovered. He strongly recommended the use of lime casein colours for artistic completion because he considered pigments bound with limewash too opaque and obviously new. Furthermore, the organic binding medium impregnated the original paint and protected it so that it could survive an aqueous cleaning in the future.[13] The quest to find an appropriate binding medium that could, when mixed with pigments and various clays, imitate the appearance of the aged original paint, and also, using only the medium in a diluted form, function as a consolidant, dominated Rothe's work and led to the development of the so-called 'Carlsberg Preparation' (a concoction consisting of alkaline soap, oil, resin, casein, turpentine, wax and camphor that was created in the laboratory of the Carlsberg brewery and used on Danish wall paintings in the period 1916–1932).[14] Despite Rothe's general acceptance of re-painting and reconstruction as a crucial element in the treatment of wall paintings, he often called attention in treatment reports to those parts of the painting that were unretouched by conservators as something especially important and unique.

Egmont Lind, Rothe's assistant and successor, has earned a place in the history of restoration for having developed, in the early 1930s, a unique retouching methodology. Lind's retouching method preceded the Italian *tratteggio* (*rigatino*), while being distinctly different in its visual effect.[15] Lind developed his approach to retouching a few years after the principle of discernible additions was published in *The Athens Charter* in 1931.[16] However, there seems to be no concrete evidence that he was actually inspired by the Athens document. The so-called basket-weave pattern, implemented for the first time in Broager Church in 1934, consisted of small groups of short parallel lines arranged in alternating directions adjacent to each other, but not overlapping (Figure 8.2). The grouping of the lines formed a patchwork pattern, reminiscent of wickerwork or woven cloth. While integrating the damaged area into the plane on which the painting was perceived, the method intentionally marked the input of the restorer, and thus shared with tratteggio a common theoretical explanation and justification. The actual design of the basket-weave pattern was inspired, just as tratteggio was, by the original painting technique of the artwork on which it was first used. Just as the parallel hatched lines on the Early Italian Primitives were interpreted and translated into the more rigid and mechanical-looking tratteggio, the imprint of bristles from a broom loaded with limewash, creating sweeping, criss-crossing brushstrokes in the ground layer of the Gothic wall paintings in Denmark was also transformed into a repetitive stylized patchwork pattern. However, this pioneering example of modern conservation practice, marking the first attempt to differentiate authentic material from later additions, did not become widespread – even Lind did not use it consistently, nor did it last more than a quarter of a century. Lind's achievements unfortunately lost ground to the old practice of artistic restoration. By the time the School of Conservation was established in Copenhagen in 1973, the aesthetic completion of wall paintings did not include an intellectually or empirically inspired retouching methodology.

Figure 8.2 *Detail from the wall paintings in Broager Church, showing the stylized 'basket-weave' retouching executed during the restoration in 1934. Photo reproduced with permission from The National Museum of Denmark.*

The predominance of material authenticity

In the course of the last quarter of the twentieth century, the attention of conservators shifted to the material substance of the paintings. This was the inevitable result of the establishment of a professional education that reached far beyond the honing of practical skills. Scientifically based diagnoses and measures have replaced remedial cosmetic interventions. The commitment to minimal intervention and preservation of the authentic material substance, coupled with a reaction against the relatively recent persistent practice of over-painting and reconstructing in the artist-restorer tradition, has swung the pendulum in favour of restraint. This attitude, which had slowly built up over the decades, has laid the foundation in Denmark for the predominant treatment of the lime-based paintings with like materials in order to restrict contamination with foreign substances and minimize alteration of physical properties. Newly uncovered wall paintings, particularly when they have survived as fragments, are now often treated as archaeological objects, where information about the technique of execution can be gleaned from a surface unadulterated by a restorer's brush. Beauty is seen in details, such as brushstrokes where the traces of bristles are imbedded in the paint, demonstrating the phenomenon of how desirable aesthetic qualities are found in objects believed to be genuine.[17] Fascination with material authenticity can, however, sometimes have the impact of a two-edged sword when it comes to presentation (viewing from a distance): The plaster repairs providing the sole aesthetic augmentation for the Romanesque wall paintings fragments (restored 1989) in Gundsømagle Church provide a neutral background focusing our attention on the remnants of the painting (Figure 8.3); on the other hand, the plaster repairs (covered with tinted limewash) on the Gothic wall paintings in Nødebo Church (restored in 1982) create visual disturbances as forms competing with the composition of the painting resulting in a pronounced subjugation of the works' artistic dimensions (Figure 8.4).

An extreme example illustrating the cultivation of material authenticity is seen in the recent conservation that took place in 1999–2001 in Vrigsted Church.[18] The renovation transformed the non-descript whitewashed interior into an archaeological display testifying to the building's rich history. Disconnected fragments of paintings from at least seven periods, from the 1100s to the 1800s, emerged in bits and pieces everywhere on the walls and the vaults when they were stripped of plaster of relatively modern date. The surface of the walls and vaults, a 'ruinous' display of authentic material substance, coupled with a new floor and a modern altar and pews, is, of course, historically inauthentic. The interior, reminiscent more of theatrical scenery than a historically authentic site, is an artificially created palimpsest (Figure 8.5). One can argue, however, that most historic church interiors are just that – palimpsests: medieval wall paintings are experienced together with church furniture from another period, often with modern

Figure 8.3 *The aesthetic enhancement of the Romanesque wall paintings executed in 1989 in Gundsømagle Church was limited to plaster repairs tinted with coloured sand. Photo reproduced with permission from R. Fortuna.*

lighting fixtures and altar tapestries. What is disconcerting in the case of Vrigsted is that this palimpsest seems to be staged, created overnight, at the stroke of a wand, which is often the case when a major interior renovation takes place. Despite (or maybe due to) its controversial character, Vrigsted Church is an interesting example of the implementation of non-conventional solutions guided by a strong aesthetic vision.

The emphasis on preserving authentic material substance can sometimes lead to highly unusual situations, as in Hedensted Church, where a wall painting was unexpectedly found in 2001 on the wall of a bricked-in doorway. The event elicited both frustration and excitement on the part of the church community, who had just had an annex built, which was to be linked to the church via the very doorway where the painting was found. Recognizing, in particular, the historic value of the newly uncovered painting, the church council paid for the time-consuming and expensive process of detachment and transfer to a movable support (epoxy/fibreglass) that was hung in another location in the interior (Figure 8.6). The surface topography of the painting was extremely uneven because boulders

Figure 8.4 *The Gothic wall paintings in the chancel in Nødebo Church were treated as archaeological objects in 1982, with no aesthetic integration of damages. Photo reproduced with permission from R. Fortuna.*

were used to fill in the doorway. In order to recreate this surface faithfully, it was recorded digitally (with a Faro arm) before the paint layer was skinned off (by *strappo* technique).[19] The bizarre element in the process was the subsequent moulding of the flexible and thoroughly plasticized paint layer onto a three-dimensional artificial support, which was then mounted on the wall in the nave in the adjacent bay to where it was found – a prepared specimen attempting to project an authentic message, but in reality only a simulacrum of dubious material authenticity. One can also question whether this object relays any other aspect of authenticity. The painting on the door was a fragment of a monumental decoration gracing the walls and vaults. Originally, it was an integral part of the surface of the interior, but now it is a three-dimensional appendage viewed on a different plane, out of context. The surface topography might be a replication of the original, but the smooth sides of the transfer, blocking the view to the underside, might impute that a slab was cut out of the thick wall, a notion refuted by common sense (the transfer is, in fact, a hollow shell weighing about 30 kg). Does the transfer itself project an aura of authenticity? Hardly. Does this display strengthen the *genus loci* of the site? No. The fascinating aspect of this object is evoked by the technical achievement of

Figure 8.5 *Vrigsted Church, view toward east, after renovation in 2001. The concept of the renovation was to expose all the historically significant layers simultaneously, in the condition they were found. Photo reproduced with permission from R. Fortuna.*

the operation, the same emotion a masterly executed fake might elicit. However, for perceptive viewers this feeling can be marred as undesirable aesthetic qualities are detected, provoked by the incongruity of a wall painting – seemingly transferred boulders and all – gracing the wall of the nave like an easel painting. This kind of disillusion often happens in cases where imitation or re-creation is suspected, probably because the deception is not deceptive enough.[20]

Figure 8.6 *The fragment of the Gothic wall painting found in the niche formed by the bricked-in door was detached, transferred to a movable support that replicated the surface configuration, and is now displayed on the wall in Hedensted Church, east of its original location. Photo reproduced with permission from R. Fortuna.*

Camouflage retouching techniques

Of course, not all wall paintings that have been uncovered in the recent decades are fragments, which generally are rendered well in an 'archaeological' presentation, with no retouching attempting to enhance the appearance. More complete paintings have also been discovered (Sigerslevvester in 1990, Nibe in 1994, Gjerrild in 2004, Torslev in 2004) – larger works with multi-figural compositions or decorative floral motifs, and geometric patterns that accent the ribs and arches of the vault. In the case of such complete compositions, it is more difficult to ignore the artistic dimensions of the work, particularly when damages can be retouched without speculation. Various techniques have been employed to integrate lacunae: vertical lines, cross-hatching lines, dots, and a mixture of all three. All of them share a common trait: they are not applied as a consistent methodology – the nature of the retouching (ratio of painted lines or dots to the interstices between them) can vary from one lacuna to another depending on the appearance of the adjacent original paint layer – the better preserved the paint layer, the more compact the retouching. The retouching does not form a repetitive pattern. It balances two objectives: to be

The Concept of Authenticity Expressed in the Treatment of Wall Paintings in Denmark

Figure 8.7 *Torslev Church, wall paintings on the west bays of the west vault in the nave, after restoration in 2004. The reconstructed portions of the painting were executed with cross-hatched lines, striving for a fainter tone than the original fragments, but which allowed for the entire decoration to be perceived on the same visual plane. Photo reproduced with permission from Isabelle Brajer.*

recognizable as non-original input at close distance, and to avoid imposing an overall graphic matrix that could easily come to dominate the artistic quality of the original painting. These more or less random retouching methods are in effect camouflage techniques – whatever the intensity of the original surviving paint layer, it plays the dominant role (Figure 8.7). The function of the discernible retouching is to draw attention away from the damage, indirectly enhancing the original, without extending the retouching over the border of the lacuna (i.e. over-painting the original paint layer). The retouching is often particularly concentrated along the edges of a lacuna, with the aim of reducing contrast between it and the surrounding original paint layer. Despite the subjective character of this integrating process, entailing a certain degree of interpretation and requiring aesthetic sensitivity on the part of the restorer in the accentuation of the artistic features,[21] it is also employed to highlight the material authenticity of the work. In this respect it is an example of respecting Brandi's ideas of the artistic and historic bipolarities of a work of art.[22] What is of little relevance in the camouflaging process is respect for the original objectives of the painter – for example, the so-called 'artist's intent' regarding the intensity of the colours originally used. The aim of the aesthetic intervention is to present the painting in the condition it was found, but without the most visually disturbing damage, achieving, in effect, a condition of verisimilitude (Figure 8.8).[23]

Figure 8.8 *Detail of the wall paintings in Vallensbæk Church showing the reconstructed areas intentionally executed with crude brushstrokes in order to achieve the same visual effect as the damaged original paint layer. Photo reproduced with permission from Isabelle Brajer.*

The problem of authenticity in re-restorations

The decision to highlight the authenticity of form, design, materials or substance through aesthetic manipulation, or lack thereof is relatively straightforward when dealing with newly uncovered works of art. In the case of re-restorations, a standpoint often has to be taken about multiple historical treatments, creating much more complex problems. The importance of preserving historical treatments is undisputed,[24] and the drastic interventions of de-restoration are generally on the retreat all over Europe. But there are situations when it is not possible to preserve all the historical layers: improper materials causing harm to the original have been employed in the past (e.g. cement plaster repairs); historic additions can also age differently, causing unintended and undesirable visual disturbances (e.g. discoloured retouching); non-permanent retouching materials, such as dry pigments mixed with water or limewater, are lost during later cleaning processes.

The aesthetic processing in cases involving the removal (in part or in whole) of earlier restorations can be particularly difficult because the conservator is primarily interested in presenting truth, but in the process may lose track of it or may have to settle for a compromise. Such was the dilemma in Tirsted Church, where about

50 per cent of the non-original pictorial contents were removed when the Gothic wall paintings were re-restored for the fourth time in 2000.[25] In the eyes of the conservators, the removal of highly conjectural reconstructions from the past, which had been executed on repairs with hard and non-porous surfaces, strengthened the authenticity of the paintings. To the church community the meagre display of the remaining genuine fragments had nothing to do with their perception of the 'real' painting. The solution to replace the non-genuine pictorial contents with a monochromatic line was a compromise that attempted to be both truthful and aesthetically acceptable. This arguable degree of truthfulness was restricted, of course, to the reconstruction of the iconographic message, which was truthfully displayed as non-authentic.

Possibly even more problematic are cases where conservators are confronted with so substantially reconstructed and re-painted works that they are not fully sure of the extent and location of the original material, as in Fjenneslev Church in 2003.[26] The Romanesque wall paintings (dated 1125–1150), depicting historical figures – patrons that undoubtedly financed the decoration, are among the most well known examples of medieval art in Denmark. At the time of the most recent (sixth) restoration, the wall painting constituted a hodgepodge of overlapping repairs, reconstructions and over-paintings, presenting a picture of repeated disintegration and partial recreation. Archival photographs document the steady deterioration and loss of original details, such as facial features and drapery folds. The decision in 2003 to re-integrate new repairs with a discernible method (cross-hatching) might be seen as senseless in such a situation, prompting the question: Is there a point when the loss of genuine material is so great that one can begin to stress broader intangible values rather than the actual form and content that have survived? By stressing emotional values relating to the painting's social and cultural setting, we might be armed with an argument for reinstating the painting's former more complete appearance, recreating the lost details recorded on older photographs. Thus, the painting might be seen as a tangible 'souvenir' authenticating the viewers' experience of history,[27] a standpoint that might reflect a more oriental interpretation of the concept of authenticity.

Conclusion

This brief overview of practices regarding the aesthetic presentation of wall paintings presents various types of fidelity: to original aims, form, contents and materials, illustrating how different perceptions of authenticity have influenced, and continue to influence the way we (restorers/conservators) want viewers to 'see' the paintings that were treated. Following the most recent views on the ultimate meaning of our profession and its place in modern society, it would be wrong to view the different aspects of authenticity hierarchically. One might like to believe that a more ascetic intervention will result in a more objective conservation. But it is well known that 'pure preservation' represents an unattainable utopia.[28] Nevertheless, there are many conservators who attempt to follow this path, often considering themselves to be more 'correct' than colleagues who partake in aesthetic processing by retouching. The conscious refraining from all subjective decisions is, of course, in itself an expression of subjective behaviour. In a relatively short time, it will probably be regarded as subjective as the re-painting and unfounded reconstructions of yore. In addition, focusing only on material authenticity might lead to a dead end if the authentic message of the object is no longer understood. As David Lowenthal so aptly put it: Truth is not innate to the original material substance; 'authenticity inheres not in some founding moment, but in an entire historical palimpsest and in the very dynamics of temporal development.'[29] It is obvious that there will be nothing to contemplate in wall paintings if the appropriate measures are not implemented that will ensure the survival of the material substance. The question is whether that is enough.

Acknowledgement

The author is grateful to Ulla Kjær, Sissel Plathe, Kirsten Trampedach, and particularly to Susanne Ørum for insightful and helpful comments.

Notes

1. E. Avrami, R. Mason and M. de la Torre, *Values and Heritage Conservation Research Report* (Los Angeles: The Getty Conservation Institute, 2000) 3–11; Salvador Muñoz Viñas, *Contemporary Theory of Conservation* (Amsterdam: Elsevier Butterworth-Heinemann, 2005) 170, 195, 213–214; Isabelle Brajer, "Values and opinions of the general public on wall paintings and their restoration: A preliminary study", *Conservation and Access, Proceedings of the IIC Congress in London* (2008): 33–38.
2. Jukka Jokilehto, "Authenticity, a General Framework for the Concept", *Nara Conference on Authenticity* (ICOMOS, 1994) 32.
3. Jukka Jokilehto, "Viewpoints: The debate on authenticity", *ICCROM Newsletter*, Volume 21 (1995): 6–8.
4. Jokilehto (1994), p. 33.
5. Michael Petzet, "'In the full richness of their authenticity' – the test of authenticity and the new cult of monuments", *Nara Conference on Authenticity* (ICOMOS, 1994), p. 91.
6. Olgierd Czerner, "Communal cultural heritage in a unified Europe", *ICOMOS News*, Volume 1 (1991): 25.
7. "The Nara Document on Authenticity", *International Charters for Conservation and Restoration* (München: ICOMOS, 2004) 118–119.
8. David Lowenthal, "Criteria of Authenticity", *Conference on Authenticity in Relation to the World Heritage Convention – Preparatory Workshop, Bergen* (ICOMOS, 1994) 36.
9. David Phillips, *Exhibiting Authenticity* (Manchester: Manchester University Press, 1997) 165–196.
10. David Lowenthal, *The Past is a Foreign Country* (Cambridge: Cambridge University Press, 1985), p. 393. Jukka Jokilehto, (1994), p. 20.

11. Ulla Kjær and Poul Grinder-Hansen, *Kirkerne i Danmark II – Den protestantiske tid efter 1536* (Viborg: Boghandlerforlaget, 1989) 140.
12. Eugène-Emmanuel Viollet-le-Duc, "Restoration," *Historical and Philosophical Issues in the Conservation of Cultural Heritage* (Los Angeles: The Getty Conservation Institute, 1996) 314.
13. Eigil Rothe. (Unpublished) Notebook nr. 2 (1898): 64–67, 76–77, Archive of the National Museum of Denmark.
14. Isabelle Brajer, "The Carlsberg Preparation: An early 20th century surface treatment for wall paintings in Denmark," (in preparation).
15. Isabelle Brajer, "Theory, methodology and practice – Cesare Brandi and wall painting restoration in Denmark in the 20th century," *Theory and Practice in Conservation, Proceedings of the International Seminar, Lisbon* (2006), pp. 111–113.
16. "The Athens Charter for the Restoration of Historic Monuments (1931)," *International Charters for Conservation and Restoration*, (München: ICOMOS, 2004) 31–32.
17. Mark Sagoff, "On Restoring and Reproducing Art," *The Journal of Philisophy* Volume 9 (1978): 453–470.
18. Kirsten Trampedach, Sissel Plathe and P. Bech-Jensen, "Vrigsted Kirke – En bygningshistorik opdagelsesrejse gennem 900 år," *Nationalmuseets Arbejdsmark* (2002): 9–26.
19. Isabelle Brajer, *The Transfer of Wall Paintings – based on Danish Experience* (London: Archetype Press, 2002) 188.
20. Sagoff.
21. Isabelle Brajer, "Dilemmas in the Restoration of Wall Paintings: Conflicts between Ethics, Aesthetics, Functions and Values Illustrated by Examples from Denmark," *The Art of Restoration – Developments and Tendencies of Restoration Aesthetics in Europe* (München: ICOMOS, 2005) 122–140.
22. Cesare Brandi, *Theory of Restoration* (Firenze: Nardini Editore, 2005) 65–75.
23. Isabelle Brajer, "Authenticity and restoration of wall paintings – issues of truth and beauty," *Postprints of the International Meeting: Art, Conservation and Authenticities: Material, Concept, Context, Glasgow* (2007) Forthcoming 2009.
24. "Principles for the Preservation and Conservation–Restoration of Wall Paintings," *International Charters for Conservation and Restoration* (München: ICOMOS, 2004) 163. Jukka Jokilehto, "Authenticity in Restoration Principles and Practices," *Konservering Igår och Idag, NKF X Kongress, Finland* (1985) 19–33.
25. Isabelle Brajer and Lise Thillemann, "The Wall Paintings in Tirsted Church: Problems of Aesthetic Presentation after the Fourth Re-restoration," *ICOM Committee for Conservation Preprints, 13th Triennial Meeting, Rio de Janeiro*, 2002: 153–159.
26. Isabelle Brajer and Ida Haslund, "Questions of Authenticity after Six Re-restorations of the Wall Paintings in Fjennerslev Church," *ICOM Committee for Conservation Preprints, 14th Triennial Meeting, The Hague*, 2005: 1016–1021.
27. David Lowenthal (1985), pp. 293–295, 356.
28. Christian Baur, "Zurück zu Viollet-le-Duc?" *Vom Modern en Zum Postmodernen Denkmalkultus?* Arbeitshefte 69 Bayerischen Landesamtes für Denkmalpflege, 1994: 22.
29. David Lowenthal, "Changing Criteria of Authenticity," *Nara Conference on Authenticity* (ICOMOS, 1994) 128.

9

The Development of Principles in Paintings Conservation: Case Studies from the Restoration of Raphael's Art

Cathleen Hoeniger

This chapter will investigate the gradual establishment of restoration principles during the Early Modern period in Europe by focusing on the treatment of paintings by Raphael. The High Renaissance artist Raphael (1483–1520), who flourished under the patronage of Popes Julius II and Leo X in Rome, became even more famous after death than he had been during his lifetime. Influential theorists championed Raphael as the modern painter who had captured the spirit of the ancients most fully and whose art, therefore, should be emulated. The great value attached to Raphael's art meant that his works were typically owned by the wealthiest individuals and institutions. When they required restoration, exceptional trouble was taken. Sometimes this led to overly-ambitious treatments, while at other times new and improved standards were set. Often the treatments were documented with unusual care. The most notable advances in principled and ethical restoration occurred in institutional contexts and less so when Raphael's paintings were controlled by private patrons.

Scholars of the history of restoration have documented the rise of principles during the eighteenth and nineteenth centuries primarily with the evidence from published treatises.[1] However, restoration manuals do not necessarily capture the practice of the day. Even the most academic restorers might say one thing but do another. For example, Bartolomeo Cavaceppi, who associated closely with the brilliant classicist Johann J. Winckelmann, argued in his treatise of 1786 that the restorer should use the same materials as the original, and that he must not alter the original work of art. However, Cavaceppi carried out substantial renovations on many antique statues, in concert with the prevailing neo-classical taste for polished and complete nudes.[2]

In the German context, similar discrepancies can be seen between avant-garde treatises and the practice of restoration for private patrons. At the

Düsseldorf court of the Elector Palatine, sometime in the years 1742–1777, a skilled Parisian restorer, François-Louis Colins, was commissioned to remove the clusters of cherubim from the upper corners of Raphael's *Canigiani Holy Family*. He first attempted to scrape them away – a remarkably aggressive approach to a Raphael – and when that proved too onerous, Colins over-painted the angels with the colours of the sky.[3] Responding to invasive renovations of this kind, Christian Köstler (restorer in Heidelberg and Berlin, 1784–1851) wrote in his treatise on paintings restoration of about 1830 that the restorer must remain 'invisible.'[4] In short, the complex evidence of restoration history, which includes both written documentation and the surviving works of art, suggests that sometimes the objects themselves provide more reliable evidence of practice. Ideally, the material evidence of the physical changes wrought on the paintings should be interpreted alongside corresponding written documentation, so that together they may serve as a testament to the practice of art restoration and its principles.

In this chapter, the focus will be on three case studies from the nineteenth century in France and Italy because during this period enormous strides were taken to develop restoration principles. Prior to the late eighteenth century, few European paintings restorers were constrained by established rules of practice. The earliest recorded restoration of one of Raphael's paintings, for example, reveals a period in the history of collecting before rules of art preservation existed. Giorgio Vasari in the *Lives of the Artists* (2nd edn 1568) relates how the panel-painting of the *Madonna del Cardellino* (now: Florence, Uffizi) was broken into pieces when the suburban house in Florence of Raphael's friend, Lorenzo Nasi, collapsed due to a landslide from Monte San Giorgio in 1547. Vasari explains that pieces of the painting were found among the ruins, and that Lorenzo's son, Giovanbattista, who loved art, put the fragments back together as best he could.[5] Recent X-ray studies of the painting in the Uffizi provide evidence of how the panel was joined together along several break lines.[6] Although Vasari does not say so, one can presume that the patron sought the help of a Florentine artist in the reconstruction of his family's revered Madonna.[7]

Indeed, it was the normal practice prior to the eighteenth century and the institutional developments in France that came alongside the rise of the art museum, for painting restoration to be carried out by artists as one of their many workshop activities. Though their lengthy apprenticeship would have ensured knowledge of the making and style of panel-paintings and frescoes, the only principles that seem to have governed such early restoration work were those articulated by the patron. The price the patron was willing to pay determined the quality of the artist chosen for the commission and the value of the materials used.

Often the artists who worked as restorers during the Early Modern period were not the most gifted. Sometimes their interventions were ambitious and, by modern standards, unprincipled. Indeed, when Raphael's Cartoons – the preparatory gouaches made in 1515–1516 as templates for the weaving of the Sistine Chapel tapestries – were restored in England in the late seventeenth century, the practitioners who undertook a hazardous treatment were not prominent painters.[8] According to the diarist George Vertue (1683–1756), Raphael's Cartoons were 'repaird or joynd together in K. Williams reign. by Mr Cooke. Painter. & the Surveyor of the pictures Mr Walton father of the present Mr Walton'.[9] Both restorers were members of the royal household. Henry Cooke the Younger (1642–1700) was a history painter, and Parry Walton (d. circa 1700) was custodian and restorer of the royal picture collection. To serve as templates for weaving tapestries, the Cartoons had been cut into vertical strips early on, so they could be laid under a loom of about a metre in width. Cooke and Walton reassembled the Cartoons permanently by gluing the strips onto large canvas backgrounds.[10]

Even with the equipment and materials available today, the procedure of reassembling and lining fragile strips of gouache on paper in this way would be extremely tricky. Presumably, Cooke and Walton had little choice but to attempt the restoration requested by the King himself – William III. Principles of restoration were not in force, beyond the emphatic desire not to harm the prized art of Raphael. During the eighteenth and nineteenth centuries, moreover, private patrons continued to control the way their art works were handled, and frequently radical transformations were requested for reasons of taste and interior decoration.[11] The restorers who enacted these treatments seem to have been oblivious to the growing concern in institutional environments that one primary principle govern restoration; namely, that works of art be preserved in as close to their original form as possible.

It was at the Louvre in the years surrounding 1800 that restoration rules were first firmly established with the expressed goal of protecting valuable, original art works. Indeed, as is well known, the development of the public art museum in Europe, a development in which the Louvre led the way, strongly contributed to the establishment of a more 'modern' approach to restoration. Following the French Revolution, the Louvre, which had housed part of the art collection of the French kings, was transformed into a state museum under the revolutionary government.[12] A new museum administration began to set in place guidelines for restoration, which included dividing practitioners into specialists of structural treatments and pictorial restoration. Lines of command were established and committees appointed to supervise restoration.[13]

Although political forces led to the reopening of the Louvre as a public institution in 1789, several intellectual factors contributed to the changes that would occur there in art preservation. The rise of empiricism in chemistry and physics played a role in the advent of restoration principles that involved rationalizing the methods and ensuring correct practice. Liberal 'artistic' restorations by individuals gave way to interdisciplinary projects involving groups of experts including scientists. The careful and systematic approach that was being taken in archaeological research to layers of historical evidence also affected the way condition and restoration treatments were documented at the Louvre.

However, without the sudden arrival of a large body of valuable works of art in fragile condition, the establishment of rules of treatment might not have been taken as seriously. In the years 1796–1799, with Napoleon as leader of the French army in Italy, hundreds of Renaissance and Baroque paintings were confiscated as trophies of war from cities captured by the French. Several of Raphael's famous altarpieces were packed up and carried on ox-driven carts and by sea on frigates. They are documented on arrival in Paris as in very poor condition. Major restorations of very large panel-paintings were undertaken with some urgency and with unusual administrative care. The concern to preserve 'the original' was often paramount.[14]

The restoration carried out on Raphael's *Madonna di Foligno* in the years 1800–1801 is especially well documented and can be interpreted as a landmark in the development of principles. The altarpiece had been 'legally' taken from Foligno in 1797 as a condition of the peace treaty of Tolentino.[15] Before crating the painting for transport to Paris, the French experts appointed to accompany Napoleon and to choose the works of art to be taken, assessed the condition as very fragile and attempted to secure detaching areas of paint. Upon arrival in Paris, the painting was examined initially by Jean-Baptiste-Pierre Le Brun, the conservator in charge of restoration at the museum.[16] Le Brun described how a large crack ran down through the upper half of the painting and he noted the location and size of many areas of paint loss and detaching paint. Realizing the seriousness of the problem which faced the Louvre because Raphael's Foligno altarpiece was very famous, the administration arranged for a committee of experts to be formed, so that the condition and treatment of the painting could be managed as carefully as possible.

The result was the appointment of an interdisciplinary committee by the Institut National to supervise the restoration. The creation of such teams of experts to watch over treatments was critical for the establishment of restoration principles, since committees helped to ensure that the theories were put into practice. Supervision also ensured transparency. The administration of the Louvre had learned from the mistakes made by previous generations, who had permitted restorers to work freely and secretly. The most famous case involved Robert Picault, who transferred from panel to canvas several invaluable paintings, including Andrea del Sarto's *Charity* and Raphael's *St. Michael Altarpiece*, in the years 1749–1751. Though Picault was initially vaunted for immortalizing art against the decay of nature, within a decade many of the works began to suffer once again from flaking paint, and Picault's secret method and high fees generated suspicion. Recent scientific tests have confirmed that Picault used nitric acid vapour, percolated from the back through the pores of the wood panel, to cause the gesso layer to expand and lose its adhesion to the paint layers.[17] Because he had charged exorbitant fees for what, it was gathered even in his day, was a relatively simple procedure as far as the chemical ingredients were concerned, and because he refused to reveal his secret, Picault fell from favour by 1766, and the Louvre officials insisted in future on openness in restoration procedures.

The supervisory committee formed for Raphael's *Madonna di Foligno* is also significant because it was interdisciplinary. Two prominent painters and two famous chemists formed the committee. Though a committee of experts had been consulted in Paris in relation to the restoration of paintings in the King's collection as early as the 1740s, the participation of scientists is new in 1800–1801.[18] The roots of the practice, however, can be found in the committee of experts appointed to advise Napoleon on the works of art to be confiscated during the First Italian Campaign of 1796–1797. The 'commission' members in Italy included four scientists, as well as several painters and a sculptor. Though the scientists were present to select scientific instruments for transport to Paris, they must also have been consulted about the more technical aspects of the packing and shipping of works of art, such as how to protect the canvas paintings from the effects of moisture. One of the scientists on the Italian campaign, Claude-Louis Berthollet, was chosen subsequently as a member of the committee for Raphael's *Madonna of Foligno*. A further reason for the participation of the chemists in the Raphael restoration was to encourage restorers to adopt the materials newly developed as a result of scientific research. Indeed, Guyton de Morveau, the second chemist chosen, had carried out experiments on the permanence of artists' materials, including tests to determine which white pigments would blacken when exposed to polluted city air.[19]

The chemists Berthollet and Guyton de Morveau wrote the first part of the report of 1801, which describes the poor condition of the *Madonna di Foligno* and then explains the structural treatment carried out by François-Toussaint Hacquin.[20] They described the condition as extremely dangerous: The panel was rotten in parts, full of worm holes, and a large crack ran down through the upper half. Because the support had decayed so much, the ground and paint layers were destabilized. Extreme fluctuations in dryness and humidity due to wartime conditions and the transport of the altarpiece had contributed further to flaking paint. In addition, large areas of the image were obscured by previous re-paintings that had darkened, and by discoloured varnish. Despite an open acknowledgement of the dangers attending the precarious procedure of transfer, the committee and the administration became convinced that there was no other option but to transfer the paint layers to a new support, and to a canvas rather than a new wood panel, because of the fear that the painting might experience further fluctuations caused by changes of climate on a wood support.

The chemists went on to describe the transfer technique used by Hacquin in remarkable detail and with great clarity. Evidently, the presence of intelligent committee members ensured that the restorer explained and justified his materials and methods, thereby promoting openness, caution, the transfer of knowledge and the perfection of technical procedures that arose from critical scrutiny by experts. As the 1801 report stresses, museum surveillance has led to the best restoration methods being followed. Instead of detaching the paint layers by breaking down the gesso ground with a chemical steam and then removing the paint from the original wood support, as the controversial Robert Picault had done about fifty years earlier, Hacquin gradually planed down the wood-panel from the back to reveal the paint layers beneath the ground. The desiccated paint was treated to reduce the dryness and increase adhesion and, finally, the paint layers were applied to a canvas covered with a new ground. Hacquin added a second linen to the back of the painting to function, effectively, as a lining to the new canvas support.

In the second part of the 1801 report on the *Madonna di Foligno*, the painters on the committee, François-André Vincent and Nicolas-Antoine Taunay, emphasized the experience and sensitivity of the pictorial specialist Mathias Bartholomäus Roeser (originally from Heidelberg, 1737–1804). It was Roeser's objective to 'harmonize the restoration work with that of the master ... and to make the intervention disappear to the point that the eye ... cannot distinguish the hand of the artist-restorer from that of the master.'[21] The desire to return Raphael's paintings to their supposedly original aspect by repainting and blending in areas of loss came from a genuine belief that a painting had to be complete to be fully appreciated and that losses would mar, in a disrespectful way, the artistry of the greatest of the moderns. However, as the cleaning of the *Madonna di Foligno* in the 1950s revealed, Roeser's earlier approach had involved filling the losses to the gesso ground, and compensating with repainting to areas that were, on average, three times as large as the actual losses of original materials.[22] As will be discussed below, one of the most effective critics of such an approach to loss compensation was Giovanni Cavalcaselle, who argued that the restorer's brush should only be used inside the confines of damage to lightly integrate the losses with the conserved parts of the original.

In every era the philosophy of art preservation has been tied to broader conceptions about art, including what is considered 'original.' The principles governing paintings restoration could only change noticeably when the definition of what constituted the 'authentic' work of art was re-examined. In the period when transfers were popular in Paris from 1750–1850, the conception of the 'original' seems to have encompassed the painted or re-painted pictorial layers, but not necessarily the entire material structure of the painting. One must be cautious, however, since methods for fumigating and re-stabilizing wooden panels had not yet been perfected, and transfers were undertaken as the only means of salvaging an image which was in danger of being lost forever because the paint was detaching. An exception was made in Paris, however, when Raphael's most famous altarpiece, the *Transfiguration*, was restored in 1802 without a transfer from its unusual cherry-wood support.[23] Similar caution was given to Raphael's *Marriage of the Virgin* in Milan by a restorer whose articulated principles included the preservation of the material object as a whole.

Indeed, when the famous Milanese portrait painter and paintings restorer, Giuseppe Molteni, came to treat Raphael's *Marriage of the Virgin* in the years 1857–1858, he put great energy into salvaging the painting on its original panel. Although the official report Molteni penned upon completion of the restoration suggested that he considered the wood panel to be an integral part of the 'original, historic aspect' of Raphael's altarpiece, a close look at the circumstances reveals that Molteni chose not to transfer for practical as well as idealistic reasons.[24] These practical reasons involved, first and foremost, the quasi-public nature of the restoration of a painting considered the most valuable of any owned by the new regional gallery. Molteni served as chief paintings restorer at the Brera Gallery in Milan from 1855 on, and in 1861 he also assumed the directorship of the gallery. The Brera Pinacoteca had officially opened in 1803 as a national gallery for Lombardy, and the principles underlying the way art was to be enshrined there followed in the footsteps of the Louvre.[25] Significantly, one of the lessons learned was that committees should be set up to supervise restoration. The archival records relating to the restoration of the Raphael reveal that Molteni had to report to two separate committees, to the Brera President and to the provincial government.[26]

Molteni actually had instigated this process of consultation. Indeed, it was his express desire that restoration be further systematized at the Brera. Several letters to the Brera President record Molteni's concern that picture restoration be more regulated in order to facilitate his work and to improve the overall maintenance of the collection. Molteni's personal effort to systematize restoration at the Brera is also evident from lists he sent to the President of the conservation needs of paintings and their urgency, and from the method he established to document the condition and restoration treatments in a chart format.

It is not fully clear whether the cautious approach that Molteni adopted for the Raphael was due to pressure from committee members or his own preference for minimal intervention in the case of a masterpiece. Some documents do suggest that Molteni was being very carefully watched and that the authorities required that he work with the utmost care and using only proven methods. For example, on 19 August 1857, Molteni wrote to assure the President of his wisdom and foresight.[27] His letter reveals that two committees have been put in place, both the regular Committee on Paintings and a second one, to oversee Molteni's work and to report to the President and the Minister of the Interior. Molteni has been asked to present his treatment proposal in advance for committee approval. He is also to report regularly and accurately to the committees. Molteni tries to persuade the President that little risk is involved, citing, by way of an extreme contrast, Leonardo's *Last Supper*, 'which is . . . already . . . almost totally destroyed by early repainting and inauspicious restoration' and a painting that 'cannot be detached from the wall' for treatment, and for which a newly invented restoration method needs to be found before any treatment can be effective. The Raphael, however, is an oil painting on panel, and the methods of conservation will be carefully 'outlined and justified' and are all 'well understood.' Molteni stresses his understanding that he is taking on 'the huge responsibility of working . . . on a painting by Raphael that is . . . of such a rare freshness of colouring, that the attention of the entire artistic world will be turned [to watch him].' Molteni promises not to prejudice in any way the appreciation of the 'originality of this rare treasure.' His restoration 'will be limited to the consolidation of the colour layer, in the places where it is lifting up in the form of bowls; to the removal of the old varnish and the yellow stains; and to the levelling (or flattening out) of the wood panel.'

Though the transfer method was still very popular at this time, it is important to note that Molteni never once articulated the possibility of transferring the paint layers to a new support. Although he recognized the paint layers were subject to some lifting and that the primary cause was the undermined wood support, Molteni was not comfortable performing a transfer, in part because of his cautious nature and scrutiny by the committees, and in part because he lacked experience with the technique. The Brera could have chosen to bring in a well-known transfer expert from Florence – Giuseppe Secco-Suardo. That the practice was of interest to the Milanese is revealed by the Brera's appointment in 1867, following Molteni's death, of Antonio Zanchi, a restorer who had been specifically sent by the Ministry in Milan to study the transfer technique under Secco-Suardo in Florence.[28] However, for the Raphael, the committees seem to have been more comfortable allowing Molteni to restore the panel support in his own way. The result was a remarkably integrated treatment for its day, in which Molteni, by applying wet cloths and pressure to the back of the wood-panel returned the painting to a more planar condition.

Molteni also showed unusual restraint in the re-touching of the paint surface of the Raphael. When he restored paintings for collectors outside the context of the Brera, most importantly Sir Charles Eastlake, Molteni often made aesthetic changes to elements of compositions, supposedly in order to reassert the spirit of the original artist.[29] It seems plausible that his normally artistic approach was constrained by the Brera committees, who closely supervised the cleaning and re-touching of the Raphael.

Molteni's achievement was praised by most contemporaries and has survived the test of time since the altarpiece has not been judged in need of a major conservation since.[30] However, one voice spoke out critically in Molteni's day; namely, Giovanni Cavalcaselle, who lamented that after Molteni's restoration, the colouring of the picture was imbalanced and 'out of focus.'[31] Yet Molteni had taken, what was for his day, a cautious and consultative approach, and he worked within the framework of the public art gallery, where the audience expected famous paintings to be presented in an aesthetically unified way. By contrast, Cavalcaselle was an art-historical researcher, a writer on art, and a politician, rather than a museum employee. He was interested in the documentation and cataloguing of Italian art, and disturbed by the poor condition of many historical paintings, which he had encountered while researching all over the peninsula for the surveys he wrote together with Joseph Crowe. Cavalcaselle found that many mural cycles were virtually illegible due to poor environmental conditions and neglect, with the additional complication of heavy repainting, which obscured the authorship and even the subject matter.

Cavalcaselle used his government positions to press into law principles of minimal intervention and gradually became the most influential and forward looking advocate for restoration of the age. His essential views, that works of art should be preserved without restoration or renovation, and that repainting should be limited to what is now called 'in-painting,' within the confines of paint loss, were radical for their day. He established guidelines for the proper training of restorers, and he also fought to set up governmental supervision for the restoration of works of art outside the precincts of public museums.[32]

One case that illustrates his attempt to establish controls outside the museum involves the fresco of the *Trinity and Saints* in the Chapel of San Severo, Perugia. Raphael had painted the top two-thirds of the composition, with the depiction of the Trinity and seated saints in about 1502–1507, and a lower row of standing saints had been completed by Perugino in 1521. In 1871 Cavalcaselle tried to stop a restoration campaign on the fresco, which had been arranged without proper approval. The scaffolding had already been erected to enable Nicola Consoni from the Vatican restoration laboratory to begin his work. Consoni had completed detailed preparatory cartoons, which revealed his intention to recreate areas of Raphael's fresco that had been lost over the centuries.[33] Cavalcaselle's strong objections brought the project to a standstill for a short time, and ultimately reduced the amount of repainting that was carried out. He persuaded the Ministry in Perugia that correct protocol had not been followed and insisted that the approval come from a committee on which his was the strongest voice. The outcome, predictably, was a strong restraining order that any restoration had to be strictly limited to stabilization, in other words to consolidating the intonaco plaster. In 1875, however, Cavalcaselle and his committees were over-ruled when the powerful art inspector for Umbria brought Consoni back, though he was only allowed to carry out a limited amount of repainting.

Cavalcaselle's insistence on minimal intervention in this case was significant because the principle of restraint was not just expressed in theory, but actually had an impact on practice. When the monograph on Raphael that Cavalcaselle wrote together with Crowe was finally published in 1882, however, the reformer's dismay at the failure of the legal structures he had worked to institute was voiced in a footnote concerning the over-restoration of the Perugia fresco:

> There is no part of this painting which has not been injured ... due in part to early retouching, in part to [more recent] 'restorations'. ... [T]he result of [Professor Consoni's] operations, which would have been avoided if the municipality had attended to the instructions of the ministry which forbad all retouching or stippling with colours, is unhappily that the whole fresco is covered over with an opaque fog, which adumbrates and weakens most of the wall painting.[34]

This chapter has traced the gradual emergence of some principles of painting restoration during the eighteenth and nineteenth centuries in Europe, with a focus on the restoration of paintings by Raphael. A sequence of case studies has highlighted how, by the late nineteenth century, the work of art began to be interpreted as a historical document, in such a way that the complete material constitution of the painting came to be of potential value, and the sacrifice or covering over of original materials was criticized. It was left for the twentieth century to develop firmer, more consistent and more effective systems for the implementation and supervision of restoration principles.

Acknowledgements

The author particularly wishes to thank Emanuela Daffra (Brera, Milan) and Alan Derbyshire (Victoria and Albert Museum, London) for their kind assistance.

Notes

1. See, for example, Alessandro Conti, *Storia del restauro e della conservazione delle opere d'arte* (Milan: Electa, 1973; rev. edn 1988), chapters 7–9.
2. *Raccolta d'antiche statue, busti, bassirilievi ed altre sculture restaurate da B.C. scultore Romano* (Rome, 1786), 3 vols. See: Michelangelo Cagiano de Azevedo, *Il gusto nel restauro delle opere d'arte antiche* (Rome: Olympus, 1948) 68–70.
3. Hubertus von Sonnenburg, *Raphael in der Alten Pinakothek*, exh. cat., Munich, Alten Pinakothek (Munich: Prestel Verlag, 1983) 39–41.
4. Christian P. Köstler, *Ueber Restauration alter Oelgemälde* 3 vols (Heidelberg, 1827–1830) Volume 1, 49–50 (rpt. Leipzig: Thomas Rudi, 2001).
5. Giorgio Vasari, *Le vite de' più eccellenti pittori scultori ed architettori* (1568) Gaetano Milanesi (ed), 9 vols (Florence: G.C. Sansoni, 1878–1885) Volume 4 (1879) 321–322.
6. *Raffaello a Firenze: dipinti e disegni delle collezioni fiorentine*, exh. cat., Florence, Palazzo Pitti, 1984 (Milan: Electa, 1984) 77–87.

7. Anabel Thomas, "Restoration or Renovation: Remuneration and Expectation in Renaissance 'acconciatura'", *Studies in the History of Painting Restoration*, eds., Christine Sitwell and Sarah Staniforth (London: Archetype Publications, 1998) 1–14; and Cathleen Hoeniger, *The Renovation of Paintings in Tuscany, 1250–1500* (Cambridge and New York: Cambridge University Press, 1995), chapters 3 and 5.
8. John Shearman, *The Pictures in the Collection of Her Majesty the Queen: Raphael's Cartoons and the Tapestries for the Sistine Chapel* (Bristol: Phaidon, 1972) 138, 147–149.
9. George Vertue, *Vertue Note Books*, Vol. III [1730], The Walpole Society, Vol. 22 (Oxford: Oxford University Press, 1933–1934) 43.
10. Sharon Fermor and Alan Derbyshire, "The Raphael tapestry cartoons re-examined", *Burlington Magazine*, Volume 140 (1998): 236–250; and Joyce Plesters, "Raphael's Cartoons for the Vatican Tapestries: A Brief Report on the Materials, Technique and Condition", *The Princeton Raphael Symposium*, eds., John Shearman and Marcia B. Hall (eds) (Princeton, N.J.: Princeton University Press, 1990) 111–124.
11. Alessandro Conti (*Storia del restauro*, 1988, p. 91) mentions that at Versailles, among a long list of paintings enlarged, cut down, or changed from tondo to oval form, one entire group was enlarged to fit into decorative architectural surrounds.
12. Andrew McClellan, *Inventing the Louvre: Art, Politics, and the Origins of the Modern Museum in Eighteenth-Century Paris* (Cambridge and New York: Cambridge University Press, 1994), especially chapters 3 and 4.
13. Ann Massing, "Restoration Policy in France in the Eighteenth Century", *Studies in the History of Painting Restoration*, eds., C. Sitwell and S. Staniforth (London: Archetype, 1998) 63–84.
14. Volker Schaible stresses the importance of 'ethical and aesthetical reflections' in motivating changes in restoration policy, "Die Gemäldeübertragung: Studien zur Geschichte einer 'klassischen Restauriermethode'", *Maltechnik – Restauro*, 2/89 (April 1983): 96–129.
15. Deoclecio Redig de Campos, "La *Madonna di Foligno* di Raffaello: Note sulla sua storia e i suoi restauri", *Miscellanea Bibliothecae Hertzianae: zu Ehren von Leo Bruhns, Franz Graf, Wolff Metternich, Ludwig Schudt*, Römische Forschungen der Bibliotheca Hertziana, Bd. 16 (Munich: Anton Schroll & Co., 1961) 184–197, with an appendix of documents by Gilberte Émile-Mâle; and M.-L. Blumer, "Catalogue des peintures transportées d'Italie en France de 1796 à 1814", *Bulletin de la Société de l'Histoire de l'Art Français*, 2 fasc. (1936): 244–348, and 305–308. On the value Raphael's works held for Napoleon and his deputies, see Martin Rosenberg, "Raphael's *Transfiguration* and Napoleon's Cultural Politics", *Eighteenth-Century Studies*, Volume 19/2 (1985–1986): 180–205.
16. Gilberte Émile-Mâle, "Jean-Baptiste Pierre Le Brun (1748–1813): Son rôle dans l'histoire de la restauration des tableaux du Louvre", *Mémoires de Paris et de l'Ile de France*, Volume 8 (1956): 371–417.
17. Gilberte Émile-Mâle, "La première transposition au Louvre en 1750: *La Charité* d'Andrea del Sarto", *Revue du Louvre et des Musées de France*, Volume 3 (1982): 223–231, with technical analysis by Jean Petit.
18. The cautious approach at the Louvre is already seen in a report of 1749 concerning Raphael's *St. Michael Altarpiece*; see: Conti, 126–127.

19. Barbara W. Keyser, "Between science and craft: The case of berthollet and dyeing", *Annals of Science*, Volume 47 (1990): 213–260.
20. "Rapport a l'Institut National sur la restauration du Tableau de Raphael connu sous le nom de La Vierge de Foligno, par les citoyens Guyton, Vincent, Taunay et Berthollet", J.D. Passavant, *Raphael D'Urbin et son père Giovanni Santi* 2 vols (Paris: Renouard, 1860) Vol. 1, 622–629.
21. Passavant (1860) Vol. 1, p. 627 (my translation). See: Roger H. Marijnissen, *Dégradation, conservation et restauration de l'oeuvre d'art* (Brussels: Arcade, 1967) 41–42, on the seamless repainting of the Brussels restorer Dumesnil.
22. Redig de Campos, 190.
23. Gilberte Émile-Mâle, "La Transfiguration de Raphael: quelques documents sur son séjour à Paris (1798–1815)," *Rendiconti della Pontificia Accademia Romana*, (1960–1961): 225–236.
24. Giuseppe Molteni, "Relazione intorno alle operazioni fatte al quadro di Raffaello rappresentante lo Sposalizio di Maria Vergine", Ms. Archivio Vecchio della Soprintendenza ai Beni Artistici e Storici, Milano, II, Parte 54[I], reprinted as Appendix in C. Bertelli, P.L. De Vecchi, A. Gallone, M. Milazzo, *Lo Sposalizio della Vergine di Raffaello* (Treviglio: Grafica Furia, 1983) 76–80. On Molteni: *Giuseppe Molteni e il ritratto nella Milano romantica 1800–1867: Pittura, collezionismo, restauro, tutela*, exhib. cat., Museo Poldi Pezzoli, Milan, 2000–2001 (Milan: Skira, 2000), including Jaynie Anderson, "Molteni in corrispondenza con Giovanni Morelli. Il restauro della pittura rinascimentale a Milano nell'Ottocento", (47–57).
25. Simonetta Bedoni, "Giuseppe Bossi e Raffaello", *Raffaello e Brera* (Milan: Electa, 1984), 79–90; and Giuseppe Bossi, Letter to the Minister of the Interior, 10 June, 1804, reprinted in Giuseppe Bossi, *Scritti sulle arti*, ed. R. P. Ciardi, 2 vols. (Florence, 1982) Volume 1, 292–293.
26. The following discussion is based on documents in the Archivio Vecchio of the Soprintendenza, Milano. All transcriptions and translations are my own unless otherwise stated. See also: Annalisa Zanni, "Note su alcuni restauratori a Milano: Cavenaghi e Molteni", in *Zenale e Leonardo: Tradizione e rinnovamento della pittura lombarda*, exh. cat., Milan, Museo Poldi Pezzoli, 1982–3 (Milan: Electa, 1982) 250–253.
27. Letter of 19 August 1857, from Molteni to Brera President, Archivio Vecchio, Soprintendenza, Milano. A sequence of quotations from the letter will be included in this paragraph of my text.
28. G. Secco-Suardo, *Manuale ragionato per la parte meccanica dell'arte del ristauratore dei dipinti* (Milan, 1866; rev. edn 1927) 191–197.
29. Eastlake entrusted Molteni with over thirty paintings during the 1850s and 60s prior to their export to the National Gallery. Cecil Gould, "Eastlake and Molteni: The Ethics of Restoration", *Burlington Magazine*, Volume 116 (1974): 530–534; idem., "Lorenzo Lotto and the Double Portrait", *Saggi e Memorie di Storia dell'Arte*, Volume 5 (1966): 45 ff.; and Jill Dunkerton, "Gusto, stile e tecnica in due restauri di Giuseppe Molteni", in *Giuseppe Molteni* (2000) 77–83.
30. The only conservation since was a local repair in 1957 to damage caused by a political attack on the painting; Pierluigi De Vecchi, "Raphael Urbinas MDIIII", *Lo Sposalizio* 1983: 55–57.

31. J.A. Crowe and G.B. Cavalcaselle, *Raphael: His Life and Works, with particular reference to recently discovered records, and an exhaustive study of extant drawings and pictures*, 2 vols (London: John Murray, 1882, 1885) Vol. 1, 169.
32. G.B. Cavalcaselle, "Sulla conservazione dei monumenti ed oggetti di belle arti e sulla riforma dell'insegnamento accademico", *Rivista dei Comuni Italiani* Torino, 1863: 34–37. See the essays by Anna Chiara Tommasi, Bernardina Sani and Valter Curzi in *Giovanni Battista Cavalcaselle: conoscitore e conservatore*, Atti del convegno, ed., A.C. Tommasi (Venice: Marsilio, 1998) 23–33, 35–51, and 53–63.
33. Claudia Consoni, "Restauro conservativo e restauro integrativo: l'intervento di Nicola Consoni sull'affresco di Raffaello e Perugino in San Severo", *Ricerche di Storia dell'arte: Cavalcaselle e il dibattito sul restauro nell'Italia dell'800*, Volume 62 (1997): 24–38.
34. Crowe and Cavalcaselle, *Raphael*, Volume 1 (1882, 1885), 328.

10

A Critical Reflection on Czechoslovak Conservation-Restoration: Its Theory and Methodological Approach

Zuzana Bauerová

Introduction

Thanks to its location and rich history, Central Eastern Europe has been a crossroad for various national, religious, intellectual, and artistic influences, especially since the beginning of the twentieth century. The development of conservation–restoration principles was influenced by a diversity of interests, methodological approaches and newfound philosophical theories. Because of this, depicting the development of conservation–restoration theories and methodological approaches in this region becomes a very ambitious task. Yet, it is this particular quality that allows for possible interpretations of the theoretical basis of, and practical interventions in, conservation–restoration in the former Czechoslovakia. Information on society, historical circumstances, the conservation profession, and the treated object itself (including documentation of past treatments, and the optical, aesthetic and iconological qualities of the materials used) merge to build a colorful mosaic of Czechoslovakian conservation–restoration history influenced by a number of artistic and art historical concepts, ideas, opinions, and theories. Looking at the history of Czechoslovakia's intellectual background in the 1920s and 1930s helps to clarify the ethical and aesthetic issues that influenced decisions taken during past interventions, and contributes to further explanations of how new theories and methodological approaches came to be implemented.

This paper looks at the historical roots of recent Czech and Slovak conservation–restoration theories and practices that were influenced by the concept of Structuralism, an artistic and philosophical system that focuses on the interaction between language (text) and picture (visual sign) as a tool for assessing the variability of interpretation. Using examples of conservation–restoration theories developed between the 1920s and 1940s by the art historians active in the field of monuments preservation, I will demonstrate how this method, which has become central to the disciplines of linguistics, anthropology, social sciences, and art history, informed the multiplicity of sources that led to the development of conservation–restoration in Czechoslovakia. Structuralism, together with Surrealism and psychoanalysis, influenced the Czechoslovak avant-garde artistic movements, philosophy, art history, and cultural heritage preservation during one of the country's most intellectually flourishing periods, the First Czechoslovak Republic (1918–1938), and determined the interpretation of these cultural forms after WWII. Analyzing the intellectual milieu of the early to mid-twentieth century, I have identified similarities between Structuralist-informed Czechoslovak conservation–restoration practice and theory and European conservation–restoration methodological approaches introduced and implemented by Camilo Boito and further developed especially by Giulio Carlo Argan.[1] Czechoslovak conservation–restoration theories inspired by Structuralistic methodology had a direct influence on Czechoslovak conservation–restoration practice by introducing a concept of perceiving the monument in a new way. This meant understanding both its structural complexity and its historical and aesthetic polarities. At a theoretical level this methodological approach preceded the development of Brandi's ideas and, together with the natural sciences (which focus on material authenticity), these theories had a share in the implementation of concepts of aesthetic perception in an artistic conservation–restoration methodological approach widely recognized in the second half of the twentieth century by the state monument preservation, including conservation–restoration academia in Czechoslovakia.

This study is based on a review of artistic and philosophical systems, and conservation principles through the interpretation of written documents, publications, and archival documents.

Structuralism – a methodological approach in linguistic theory and art history

Structuralism, a linguistic theory of the sign, grew to become one of the most popular approaches in academic fields concerned with the analysis of language, culture, and society in the twentieth century. It explores the relationships between fundamental elements in language, literature, and other fields upon which some higher mental, linguistic, social, or cultural structures and structural networks are built. This theory of the linguistic sign (often organized as codes governed by explicit and implicit rules agreed upon by members of a culture or social group) was developed from the work of Swiss linguist Ferdinand de Saussure (1857–1913), in which he argued that language is a cultural phenomenon. As such, it produces meaning by a system of relationships: a network of similarities and differences. He defined the linguistic sign as a two-sided entity (a dyad) consisting of the inseparable signs called signifier (the form that a sign takes) and signified (the concept that the sign represents). The sign comprises the circuit between the signified and the signifier and as such composes signifying structures with encoded meanings. From the methodological point of view, his theory of signs as a coded access to an object focuses on the arbitrary nature of the bond between signifier and signified.

De Saussure's method of Structuralist literary analysis (which was published posthumously in 1916), together with the methodological approach of the Russian Formalists,[2] had a significant influence on Czech art history, theory and criticism[3] in the 1920s. The Structuralist methodological approach focused on the interpretation of semantic functions created by combinations of separated artistic elements (such as colors, lines, surfaces) within works of art using substitution and combination as Structuralistic operations. Its significant continuing influence on linguistics, as well as on semiotics, even after World War II, led to the development of a method of Structuralist literary analysis by the Prague Linguistic Circle (1928–1939), grouped in Prague around one of the Russian formalists, Roman Jacobson. Within this intellectual milieu, Czechoslovak Structuralism influenced not only literary theory, linguistics and philosophy, but also inspired art historiography (a study of writing of history of art) and, as a methodological tool, affected conservation–restoration theory introduced by the Czechoslovak cultural heritage preservation in the first half of the twentieth century. It was further developed in the 1950s, and became known as the 'Czechoslovak Conservation School.'

Czechoslovakia in the 1920s – literary theory, linguistics and art historiography

Prague became the capital of independent Czechoslovakia after the fall of the Habsburg monarchy in October 1918 and entered the golden decade of the 1920s, when it encountered virtually all the tendencies of contemporary art at once: Cubism, Cubo-Futurism, Expressionism and Fauvism. Czech culture absorbed influences from varied sources, including (in addition to 'old' Europe) Soviet Russia, the United States, and the new countries of Central Eastern Europe. Prague quickly became a setting for young literati, visual artists, and theorists who met each other in cafés in the labyrinthine Old Town, shaping their ideas, planning exhibitions and formulating their statements towards cultural and political events. Avant-garde activity in the 1920s was concentrated around the art group Devetsil (Karel Teige, Josef Sima, Jindrich Styrsky, Toyen and Adolf Hoffmeister), which became the focus of ideological, literary, artistic and theoretical activity for the generation born c.1900.

Among the intellectuals was Russian formalist Roman Jakobson, a friend of Vladimir Mayakovsky, Boris Pasternak and the Russian Futurists. He, together with Jan Mukarovsky[4] and other Czech and Russian scholars, founded the Prague Linguistic Circle, which pioneered a Structuralistic approach to language and implemented it in other areas, including the social sciences, ethnology, art, and art history. Temporally and philosophically, it appeared between two other methodological schools – Russian Formalism and the Structuralism of Neue Wiener Schule der Kunstgeschichte. While Neue Wiener Schule was based on the idea of Gestaltpsychologie and intuitive substantionalism,[5] and interpreted a work of art as an individual and unmediated entity independent from its reception, Prague-based Structuralism was based on analytical functionalism, which understood the work of art as 'a product' of social-historical reception – that is, a semiotic sign with a social function.[6] Furthermore, in Vienna, an artwork's reception was believed to be strictly subjective, while in Prague, it embodied objectivity and plurality. Czechoslovak Structuralism concentrated on relationships, complexity and the new quality of individuality and, as such, connected principles of dialectics (principles of dialectic tensions), semiotics (the concepts of signs), and functionalism (interpreting art as functional instrument).[7]

Originally, Structuralism was applied in the fields of linguistics and literal science. It was Jan Mukarovsky who generalized these theories, thereby creating a new, suitable tool for aesthetics and general art theory, using examples from contemporary visual arts – especially from artists around Devetsil (Sima, Toyen). Structuralism offered a way to concentrate on individuality, on those features that differentiate a specific object from a notion of universality. Prague's conception of Structuralism created an attractive and inspirational scientific approach that moved from description to analysis, from analysis to synthesis, and finally to a construction of the art work as an individual entity within wider social and historical contexts.[8] However, the 'users' of this methodological approach belonged to different generations, and never created a platform based on a shared methodological position. Although Structuralism played an important role in art history and theory, it remained an ideal model, introducing interpretation of the relationship between art and society as a methodological tool.[9] Since art historiography and theory had a significant impact on the conservation–restoration methodological approach of the time, Structuralism as a methodological tool turned into one of the most important inspirations for individual conservators–restorers in their effort to introduce common principles to conservation–restoration practice.

Structuralism as a phenomenon in cultural heritage preservation

Structuralist principles were applied within conservation–restoration theories by the 1930s by laying emphasis upon the relationship between aesthetical principles, art and society. In his book entitled *Artwork and Its Preservation*,[10] art historian and conservator Vaclav Wagner introduced a conservation–restoration approach based on aesthetic presentation and an obligation to restore the artistic impression of the monument or work of art, which became known as the

'Synthetic method of monument preservation.' The Synthetic method integrated object analysis with knowledge and judgments in order to present the work of art in its complexity, as one artistic unit, as a 'living art' connected with contemporary circumstances. As a consequence, every conservation–restoration intervention on a monument or work of art had to result from the evaluation of its historical and aesthetic values, interpreting them within present aims and aesthetic and artistic qualities. One of the most famous examples of this approach (lately widely criticized by Wagner's opponents) is the reconstruction of the baroque Cernin palace at Hradcany (Prague castle) that was theoretically guided by Wagner and carried out by a famous modernist architect, Pavel Janak, in 1928–1934. Modern interventions, sensitive to the original structure, were done in order to improve the architecture for the modern use.

Wagner emphasized perception, and supported concepts that stressed the artistic appearance of a monument or work of art. In his book he published several examples of conservation–restoration treatments of the day, demonstrating the difference between Analytical and Synthetic methods: a 'drastic example of the documentary tendencies' and 'restitution of the original state respecting artistic effect' respectively. The Synthetic method presented the monument, 'as the work of art valuable for (its) aesthetic effect – [since a monument presents] not only the age value, but also the quality of artwork that stems from the self-referential unity, structure, organism.'[11] In practice, this theory employed restitution of the original appearance of the monument and allowed imitation of original parts and in some cases even further integration of copies to achieve integrity and unity and a unifying artistic impression in order to suppress all disturbances of the work's appearance.

Conservator Vaclav Wagner based his theory on Jan Mukarovsky's methodological approach. In his conservation–restoration theory he quoted from Mukarovsky's famous essay, 'Can Aesthetic Value have Universal Relevance?' (1941), and pointed out the relativism of historical changes. His methodology conceptualized spiritual and physical contact between human beings and works of art as it was interpreted by a sociological approach implemented in art criticism of that time: 'It is not possible to comprehend the art only as evolution . . . There is a living human being, or even more – pure humanity, hidden behind each artwork. To find this man, to touch the endless and common humanity, that is the real goal of the art historian.'[12] In other words, the role of the art historian is to revive the work of art to the contemporary viewer through interpretation. Like the aesthetician Benedetto Croce, Wagner stressed that this interpretation relies upon intuition. He therefore enriched the Czechoslovak intellectual milieu with a combination of Croce's neo-Kantian theory and Jan Mukarovsky's Structuralism, stating: '[The] work of art acts permanently, unless we disturb its formal structure (unity, organism), or even its artistic values that correspond to [the] anthropological constant of a human being.'[13] With his emphasis on unity and complexity, Wagner accepted Structuralism's challenge and leaned towards a holistic approach to conservation–restoration. However, as Czech art historian Ivo Hlobil noted in 1985,[14] Wagner implemented Structuralism predominantly at a terminological, rather than a practical, level since he did not evaluate such topics as aesthetic value, modern art, the role of aesthetics, the artist's intent, and the essence of art in his theories of monument preservation, and by using terms such as integrity, structure, living form, and aesthetic function.

However, Wagner was not isolated in his interest in Structuralist theory. Another Czech art historian, Karel Sourek, introduced his Structuralism-influenced conservation–restoration theory within his concepts for the organization and function of the Slovak National Gallery in Bratislava during his temporary directorship in 1949–1950.[15] He understood responsibility for conservation as one of the gallery's functions and focused on the importance of a clear definition of the term 'monument.' In his interpretation, the monument of historical value was presented as a system of values as introduced by Alois Riegl in his 1903 work 'The Modern Cult of Monuments: Its Essence and Its Development.'[16] (Values related to modern conservation–restoration distinguish between memorial and present day values; memorial values include age value, historical value, deliberate commemorative value, while present day values include use value, art value, newness value, relative art value).[17] At the same time, the monument had spiritual value (respect for the past and cultural sensitivity) as introduced by Max Dvorak in his guidelines for conservation–restoration entitled the 'Katechismus der Denkmalpflege' in 1916.[18] On the other hand, he stressed the importance of recognizing the work of art as evolutionary, unlimited, and an opening into the 'creation of one spirit, one epoch and one artistic genius.'[19] For Sourek, the reason for preservation sprang from the conflict between his understanding of 'a monument' and unlimited and opened 'artistic creation' as described above. His search for aesthetic functions in a monument's unity was based on understanding a monument's 'irreplaceable and unrepeatable' values through its social context (historical value). He overcame the main conservation–restoration conflict between a monument's substantiality and its spirituality thanks to synthesis, interpreted as a restitution of a disturbed balance of functions. This methodical approach of synthesis, focusing on social and historical contexts and applying the concept of coherence between function and structure, directly refers to the methodological approaches of Jan Mukarovsky.[20] Sourek argued that conservation–restoration interventions, according to 'modern Structuralist aesthetic theories,'[21] should respect the idea of unity and the 'artistic individuality of a work of art.'[22]

The importance of his concept of principles and mission of the monument's preservation as 'applied science of theoretical knowledge of art history, art historiography and methods of auxiliary sciences'[23] stems from an exact definition of terminology and reasons for conservation–restoration intervention. With his methodological approach focused on the monument's functions (both original and new) and critical reflection of the aesthetic value during the evaluation of a monument or art work, he referred directly[24] to brisk contemporary (1940s) polemics about Analytical (represented by Czech art historian Zdenek Wirth) and Synthetic (introduced by Vaclav Wagner)[25] methods in conservation–restoration. The main argument in this polemic was the approach to the monument or work of art during conservation–restoration intervention: while the analytical method gave priority to the monument's historical value, emphasizing its position as a historical document with preferred age value, the Synthetic method focused on the restoration of the artistic aesthetic expression. Sourek saw the link between the two approaches in respecting the monument's unique complexity and unity based on its historical value and its artistic individuality.[26] Unfortunately, due to the brevity of his position as temporary director of the Slovak National Gallery in Bratislava, his conservation–restoration method, together with his progressive ideas related to the functioning of the gallery, were never implemented into the museum's practice.

The Synthetic method of monument preservation faced strong criticism in Czechoslovakia, especially from the followers of the Vienna School (Vojtech Birnbaum, Vaclav Richter)[27] and from the theoretical leader and founder of the modern state administration in the field of monument preservation in Czechoslovakia: Wagner's colleague Zdenek Wirth. As a critic of the post-war reconstruction of Warsaw and Gdansk, he deemed the Synthetic conservation method a failure because of its insufficient acceptance of modern (and, at that time, influence by functionalism) architectural interventions.[28] Moreover, shortly after 1951, political circumstances that resulted from the Communist coup in 1948 forced the main leaders of Structuralism (Mukarovsky; Bakos and Povazan in Slovakia) to dismiss publicly their respect for Structuralism and Formalism as methodological tools that were interpreted by the official Marxist ideologists as inconsistent with historical materialism. Consequently, the Synthetic method in conservation–restoration (which was accused of having a close relationship with Nazi politics by the official representatives of the state care of monuments!)[29] was rejected. This meant also that the conservation–restoration principles inspired by Structuralistic theories were not fully implemented into practice or acknowledged. Czech art historian Ivo Hlobil concludes in his 1988 study on the theories of protected town zones 1900–1975: 'Real theoretical reflection on these theories never happened. This also can be considered as a sign of pragmatism that has since prevailed in monument preservation.'[30]

The aesthetic approach in conservation–restoration

While conservation–restoration principles and theories were influenced by the methodological and ethical approaches created within the political, cultural and social situation in Czechoslovakia between the 1920s and 1940s, they were also enriched by the theory introduced within the National Gallery in Prague by its director Vincenc Kramar and the gallery's conservator–restorer Bohuslav Slansky.[31] Kramar was an art historian, who formed a unique personal collection of Cubism, and also helped to establish a modern conservation–restoration studio in the gallery, including equipment for scientific investigations. He published his expert opinions on conservation–restoration of art works from the gallery's collection[32] and, in 1937, initiated a tradition of conservation–restoration exhibitions. At the end of the 1920s, he supported the young conservator–restorer Bohuslav Slansky in his studies in leading European conservation–restoration studios (Munich, Vienna and Haarlem).[33] During these internships it was probably Max Doerner, his professor at the Academy of Fine Arts and Design in Munich (1929), who had the biggest influence on the young Slansky. Translations of the passages from Doerner's book *The Materials of the Artist and Their Use in Painting*[34] that appeared in Slansky's articles published just after his return from his studies,[35] as well as in Slansky's two-volume book entitled *Techniques of Painting*,[36] introduced to Czechoslovakia scientific methods of conservation–restoration. The interdisciplinary cooperation between the gallery's director and its conservator–restorer (Slansky was employed by the gallery from 1934) supported implementation of the aesthetic conservation–restoration methodological approach based on the critical method of conservation–restoration respecting the material authenticity of a work of art (scientific conservation–restoration), and the aesthetic conception provided by the original form structure of the whole art work (the revival of visual perception).[37] This approach allowed for retouching as an improvement of the artistic values of the treated object.

Inspired by Structuralism and Phenomenology, Slansky further developed concepts of aesthetic perception into an artistic conservation–restoration approach that he introduced in 1953 and 1956 in his *Techniques of Painting*:

> …[the] Restorer has to present well-trained craft skills, as well as [a] scientific approach, but the most important [skill] is his/her artistic feeling that allows him/her to approach the work of art with all creative understanding. Therefore, it is only this ability that allows the restorer to choose the proper conservation–restoration quality and the limits of treatment in order to complete especially the aesthetic and stylistic values of treated objects. It is not enough to be only a scientist, a historian or a hand-crafter. But all the professions have to be linked together with creative talent…[38]

Conclusions

Introduction to the flourishing intellectual background of the First Czechoslovak Republic (1918–1938) shows its close relation with cultural heritage preservation. This allows a re-evaluation of the conservation–restoration methods that influenced contemporary methodological approaches, and acknowledgement of its precedence over Brandi's ideas. Focus on hitherto marginalized (philosophical and methodological) approaches introduces conservation–restoration theory and principles of the 'Czechoslovak Conservation School' from a new, theoretically oriented, perspective. Although the Structuralist-informed methodological approach to conservation–restoration that was developed in Czechoslovakia in the first half of the twentieth century remained predominantly at the level of terminology, it directly influenced the aesthetic conservation–restoration approach widely recognized during the second half of the twentieth century generally ascribed to Bohuslav Slansky. Reviewing the artistic and philosophical systems introduced in some Central Eastern European countries offers a means to critically reflect upon contemporary conservation–restoration principles and theories of conservation–restoration in the former Czechoslovakia.

Notes

1. The analysis of the Structuralist-informed Czechoslovak conservation–restoration practice and theory points out especially philologically-inspired concepts of interpretation that can be compared with conservation–restoration methods of the late nineteenth and twentieth centuries, with the criteria and methods of Italian conservators Camilo Boito and Giulio Carlo Argan. At the end of the nineteenth century, it was Boito who introduced the critieria for intervention in conservation–restoration that referred to so-called philological restoration based on the idea of respecting the monument as a document. This philological approach, based on a philological survey of the work of art ('reading' the monument that allows rediscovery and display of the original 'text' of the object), was further developed by Argan in the late 1930s. This methodology for treating a work of art signalled an important shift in conservation–restoration activity from an artistic to a critical sphere, and as such enlarged the basis of modern conservation–retoration theory.
2. Russian Formalists (Russian Formalists School, Moscow Linguistic Circle) were the group of young Russian scholars (J. Tynanov, R. Jakobson, V. Sklovskij, L. Jakubinskij, B. Tomasevskij, B. Ejchenbaum, V. Propp, V. Zirmunskij) predominantly linguistically oriented, who during and after the Revolution introduced the principle of interpreting the literary work (especially the properties of its language, referred to as its 'literariness') as a structure (phonetic, syntactic, semantic, etc.), rather than through the contexts of its creation (biographical, historical or intellectual) or reception. The Russian Formalists radically confronted content and form in order to solve the problem of the fundamental principle of the literary work. (Formalism in art theory focuses on the form of the art work – the way it is made, its visual aspects, composition (colour, line, shape, texture), rather than on the historical background or life of the artist). Their approach was critical of all trends of literary studies of that time: Symbolist, Positivist, psychoanalytic, and sociological.
3. Structuralistic analysis of the work of art, together with phenomenology, psychology, the sociology of art, and Formalism represent the main interpretational perspectives

of art history and theory. It is based on the idea of the artwork as a structure and introduces analysis of forms, meanings and receptions. Alois Riegl (1858–1905), on the verge of the nineteenth century, introduced the idea of meaning as a structured set of codes. Russian literal critics referred in the 1910s and early-1920s to Riegl's theory. See: Hal Foster, Rosalind Krauss, Yve-Alain Bois, Benjamin H.D. Buchloh (eds), *Umeni po roce 1900* (Prague: Slovart, 2007), pp. 32–39 [Czech edition of: Hal Foster, Rosalind Krauss, Yve-Alain Bois, Benjamin H.D. Buchloh (eds), *Art Since 1900* (London: Thames & Hudson Ltd, 2004)].
4. Jan Mukarovsky (1891–1975), Czech literary theoretic and aesthetician, is best known for his association with early Structuralism and the Prague Linguistic Circle for his development of the ideas of Russian Formalism, and for a profound influence on Structuralist theory of literature comparable to that of Roman Jacobson.
5. Jan Bakos, *Styri trasy metodologie dejin umenia* (Bratislava: Veda, 2000) 168. Bakos explains the differences between the approaches towards a work of art of the Prague Linguistic Circle and Neue Wiener Schule der Kunstgeschichte as: 'While Czechoslovak structuralists understood the perception of the work of art from the sociological perspective, as a product of the particular socio-historical reception (and furthermore as a semiotic sign with social functions), art historians with the orientation towards Gestaltpsychology centered around Kunstwissenschaftliche Forschungen [Neue Wiener Schule der Kunstgeschichte] interpreted the work of art from the perspective of substantionalism (as the entity, that is independent from any reception, and is, so to speak, self-sufficient).'
6. Bakos, pp. 163–168.
7. Bakos, p. 163.
8. Mojmír Grygar, *Terminologicky slovnik ceskeho strukturalismu* (Brno, 1999) 17.
9. Bakos, p. 184.
10. Vaclav Wagner, *Umelecke dilo minulosti a jeho ochrana*, (Praha, 1946). More recently: Vaclav Wagner: *Umelecke dilo minulosti a jeho ochrana*, (Narodni pamatkovy ustav: Praha, 2005).
11. Wagner, p. 35.
12. V.V. "Stech, Cena za stari," (*Staleta Praha*, III, 1967) 10, also quoted by Ivo Hlobil, "Teorie mestskych pamatkovych rezervaci (1900–1975)," *Umenovedni studie VI* (Praha, 1985) 25.
13. Wagner, p. 35.
14. Hlobil, p. 83.
15. Zuzana Bauerova, "Teoreticke a metodicke zaklady konzervovania restaurovania v Slovenskej narodnej galerii," *Galeria – Rocenka SNG 2002*, ed., Dusan Buran, Martin Vanco (Bratislava, 2002) 185–203.
16. Alois Riegl, *Der moderne Denkmalkultus: Sein Wesen und seine Entstehung* (Wien, 1903).
17. For English version of 'values' description, see also: Alois Riegl, "The Modern Cult of Monuments: Its Essence and Its Development," in: *Historical and Philosophical Issues in the Conservation of Cultural Heritage*, eds., Nicholas Stanley Price, M. Kirby Talley Jr., Alessandra Melucco Vaccaro (GCI Los Angeles, 1996) 69–83.
18. Max Dvorak, *Katechismus der Denkmalpflege* (Wien, 1918).

19. Karel Sourek, *Co s pamatkami* (Praha: Vysehrad, 1941) 38.
20. Jan Mukarovsky, "Strukturalismus v estetice a ve vede a literature, 1940, 1941," in: Jan Mukarovsky, *Studie I.* (Ed. Miroslav Cervenka – Milan Jankovic, Brno: Horst, 2000) 9–25; Jan Mukarovsky, "Pojem celku v teorii umeni, 1945," in: Mukarovsky 39–49; Jan Mukarovsky, "Esteticka funkce, norma a hodnota jako socialni fakty, 1936," in: Mukarovsky, pp. 81–148. For an English translation from his book *Aesthetic Function: Norm and Value as Social Facts*, see also: Charles Harrison and Paul Wood eds., "Art in Theory 1900–2000," *An Anthology of Changing Ideas* Blackwell Publishing: 2002) 518–520.
21. Sourek, p. 26.
22. Sourek, p. 29.
23. Sourek, p. 26.
24. Sourek, pp. 26–33.
25. Vaclav Wagner, *Analysa a syntesa v ochrane uměleckých pamatek* (*Volne smery*, 1940), remarque 12.
26. Sourek, p. 29.
27. Vaclav Richter, rec.V. Wagner Umelecke dilo... in: *Nase veda XXIV*, 1946: 182.
28. Among his criticized conservation–restoration interventions is, for example, the reconstruction of Cernin Palace in Prague, Hradcany (done in 1928–1934 by Czech architect Pavel Janak, student of Oto Wagner), with its implementation of modern elements in an historic structure.
29. Zdenek Wirth, "Vyvoj zasad a praxe ochrany pamatek v obdobi 1800–1959," in: *Umeni V*, 1957, p. 105. See also, Ivo Hlobil, "Pamatkova pece bez teorie je non-sens," *Bulletin, Sekce pamatkove pece vedouciho pracoviste vedeckotechnickeho rozvoje*, SUPPOP, V (1988) 28.
30. Hlobil, p. 29.
31. Bohuslav Slansky (1900–1980), Czech conservator–restorer, founder of the modern conservation–restoration in Czechoslovakia and of the formal conservation–restoration education in Prague (at the Academy of Fine Arts and Design in Prague in 1946–47). Thanks to Bohuslav Slansky, Czechoslovak conservation–restoration of the 1930s, 1940s and 1950s was informed by the decisions taken and the theories implemented by foreign institutions (Doerner-Institut, the Louvre, the National Gallery in London, and Istituto Centrale del Restauro) and leading international professional associations. See: Zuzana Bauerova, "Aesthetical and Ethical Issues of Conservation in Central Eastern Europe: museum, ideology, society and conservation (Case study: Czechoslovakia 1918–1960s," (in: *Newsletter 12*, ICOM-CC Theory and History Working Group, April 2005) 11–17.
32. Vincenc Kramar, "Budoucnost obrazarny Spolecnosti vlasteneckych pratel umeni v Cechach," (in: *Umeni I*, Praha, 1918–1921) 371–394; Vincenc Kramar: *O obrazech a galeriich* (Praha, 1983) 372–373.
33. As documented for example in the correspondence between Kramar and Slansky in: Archive National Gallery in Prague, AA 2945/477/1, fond 26.
34. Max Doerner, *The Materials of the Artist and Their Use in Painting*, trans. Eugen Neuhaus, revised edition, First Harvest edition 1984. The book was first published in 1921, and until 1933 it had four German editions. The book contents Doerner's

lectures from the Academy and became the main textbook at many conservation–restoration schools.

35. Bohuslav Slansky, "O restaurovani obrazu," (in: *Umeni IV*, Praha, 1931) 174; also Bohuslav Slansky: "Zkoumani obrazu prirodovedeckymi metodamii," (*Umeni V*, Praha, 1932) 373–374.
36. Bohuslav Slansky, *Technika malby, Dil I., Malirsky a konzervacni material* (Praha, 1953), 51, p. 52.
37. Vratislav Nejedly, "Zur entwicklung der Gemälderetousche in den Tschechischen Ländern im 20. Jahrhundert," (*Die Kunst der restaurierung, Entwisscklungen und Tendenzen der Restaurierungsästhetik in Europa*, ICOMOS Hefte des Deutschen Nationalkomitees XXXX, ed. Urschula Schädler-Saub (München, 2003) 259–268. More on Kramar and his approach can be found in Ivo Hlobil, "The Reception and First Criticism of Alois Riegl in the Czech Protection of Historical Monuments," in: *Framing Formalism, Riegl's Work*, ed., Richard Woodfield OPA- G + B Arts International Imprint, 2001) 183–194.
38. Slansky, p. 154.

11

The Problem of Patina: Thoughts on Changing Attitudes to Old and New Things

Helen Clifford

The intention of this paper is to encourage a broader consideration of the culturally specific values and contexts we place upon 'old' things. How deeply do we think about the different values of an object within the culture that made it, that used, kept and collected it? Is it possible to identify changes in attitude to objects as they aged? As an historian of material culture I am fascinated by the relationship between objects and words, how things are described, and how their value (in the broadest possible sense, economically, culturally and socially) changes over time. We will enter this rich terrain with a particular focus on wrought silver, made in England in the eighteenth century, a period of fundamental change in manufacture and consumption, which heralded the birth of 'modern Britain.'

There are four specific 'arenas' of consideration: material, workmanship, technology and consumption. After the foundation of the Bank of England in 1694, the value of silver as a hoardable reserve of convertible currency (where wrought metal could be melted and turned into cash) began to decline, as credit was secured on less tangible assets, such as stocks and shares.[1] Its value in weight, it could be argued, began to mean less than its 'look,' its fashion. Here 'fashion' means both its making, and its shape and decoration. Silversmiths' bills show that, up until the latter half of the eighteenth century, the silver content cost more than the workmanship, but thereafter began to be outweighed (literally) by the cost of its construction and decoration.

By the mid-eighteenth century, new machines, like the flatting mill, enabled sheet silver to be rolled quickly and thinly to standard gauges. The use of lighter weight metal meant that domestic objects could be made from less silver, while adopting the most fashionable forms. Their relative cost declined, making silverwares available to a wider and growing lower middle-class market.

These 'facts' about changes in the manufacture and consumption of silver have been linked to a wider phenomenon in the world of goods. The cultural historian Grant McCracken has suggested that the eighteenth century was also the time that 'Patina' ceased to be valued, as it was replaced by 'Fashion.' Patina and

Fashion, for McCracken, represent old and new systems of consumption. In the past, objects that had the wear of time, 'reassured an observer that ... [they] had been a possession of the family for several generations, and that this family was, therefore, no newcomer to its present social standing.' McCracken argues that as the attraction of age and tradition dwindled, new and fashionable goods became more desirable than those that suggested long-standing wealth and prestige.[2] These could be purchased by anyone with money. While the allure of 'patina' did not completely die, he argues that its modern manifestation 'is a pale version of its former self', and is now 'a status strategy used by the very rich alone.'[3]

McCracken, however, leaves some key questions unanswered. Across what ranks of society did this change take place, was it uniform, and was it such an uncomplicated and total shift in attitude? An investigation into the different understandings of what 'patina' is provides a means of testing his theory and linking what appears to be a physical phenomenon well known to collectors and conservators with broader issues that connect economic, social, cultural, and even moral attitudes to material culture.

First, what do we mean by 'patina?' The Latin word 'patina' relates to a type of shallow dish. Its association with the alteration of the surface of things only appeared in the 1740s.[4] It acquired a new meaning at precisely the time McCracken identifies a shift in attitude to consumption, and from then is a word that 'is as rich as the objects it describes.'[5] Patina can be most broadly defined as the weathering or aging of the exposed surface of a material, which can involve colour change: copper turns green, lead goes from silver to grey, and silver acquires a lustrous bluish surface. Patina can be created naturally by the oxidizing effect of the atmosphere or weather. However, 'the characteristic mellow lambency' of patinated silver has its origin, in large measure, in its 'physical wear-and-tear.'[6] Deeper examination of the meanings associated with the word 'patina' leads us into more complex territory.

Patina is 'everything that happens to an object over the course of time.'[7] Patina, from the eighteenth century, for collectors, increased the value of things. It was at precisely the same time that patina was supposedly losing its hold that the 'antique' came to be newly valued. An example of this valuing of the antique, as physically manifested through patina, appears in a letter of 1775 in which 'the crust or patina' of a Roman coin was directed 'not be removed [as] it is evidence of the coin's antiquity.'[8] Yet in the language of early modern retailing, the 'antique' referred more commonly to new commodities that copied shapes and decorative details of 'old' objects, most popularly in the second half of the eighteenth century from Greek and Roman architecture.

Patina is endowed with both positive and negative meanings. 'The soft, indefinite appearance of the patinated surface' has been described by collectors (who began to collect silver from the mid-nineteenth century) as 'velvet' that contrasts with the 'whiter, mirror-like surface' of modern pieces. There is embedded in this juxtaposition a moral judgment that the lustre and depth of patina is more than skin deep. The eighteenth-century blue-stocking Lady Mary Wortley Montague (1689–1762) made a distinction between the 'lustre of real worth and mere conspicuousness' in her discussion of the moral standing of women.[9] However, the contrast is not simply between age-old and shiny new. Patina itself could be viewed negatively, when a 'false patina,' artificially created, deceived the eye of the undiscriminating observer. To appreciate true from false patina is to be able to read the history of things. A body of connoisseurial literature was beginning to appear in the eighteenth century that helped the observer discriminate between the real and artificial.

The condition of silver is frequently connected with the dignity of its owners. The artist Benjamin Robert Haydon (1786–1846) complained in his diary that the servants had not cleaned his plate, and noted that with 'perspiration and violent Effort' he polished it himself, '& felt my dignity revive.'[10] From household account books, bills and receipts we can see that eighteenth-century owners of silver went to great efforts to keep their silver polished, which added to its patina. The orders for 'boyling and burnishing'[11] silver outweigh the purchase of new. Thomas Twining (1734–1804) refers to the noble families who had their plate 'cleaned & brushed every week.'[12]

Yet there is also evidence of the growing attractions of the 'new and shiny' in the eighteenth century. In the late 1770s, a new type of engraving appeared called 'bright cut.' Using the same hand-powered burins used for earlier engraving, bright cutting involved the cutting out of triangular notches of silver from the surface, which reflected the light in a manner more showy than the former cursive type of engraving. It was not just precious metalwares that catered for this new taste. Cut steel jewellery became the vogue from the 1750s; the tiny facets of metal refracted light with a glare that imitated, and for a time challenged, the popularity of the diamond, either real or paste. So, at the very time that McCracken argues the age of things is becoming less important, we have an expanding vocabulary to identify and describe and analyse it. Patina turns out not only to be a physical property of 'things,' but also a means to understand their rich economic, social, and cultural context. Where objective scientific and subjective social historical interests intersect we can be sure to be on fruitful, if often contentious, ground. In order to understand and appreciate the complexity of 'things' we need far more interdisciplinary discussion, within and beyond the museum environment. Conservators, curators and historians need to be given the opportunity to exchange information and share viewpoints. The result would be a far more subtle, if complex, appreciation of the objects that occupy them.

Notes

1. Helen Clifford, "Of consuming cares: Attitudes to silver in the eighteenth century", *The Silver Society Journal*, Autumn 2000: 53.

2. Grant David McCracken, *Culture and Consumption: New Approaches to the Symbolic Character of Consumer Goods and Activities* (Indiana: Indiana University Press, 1990) 37.
3. McCracken, p. 37.
4. C.T. Onions, rev. & ed., *The Shorter Oxford English Dictionary*, vol. II (London: Book Club Associates, 1983) 1528.
5. "Antiques Speak", *Antiques Roadshow Online*, 1 May 2006 (http://www.pbs.org/wgbh/roadshow/glossary/).
6. Seymour Rabinovitch, "The Patina of Antique Silver: A Scientific Appraisal", *The Silver Society Journal* Winter 1990: 13–22.
7. "Antiques Speak", *Antiques Roadshow Online*, 1 May 2006 (http://www.pbs.org/wgbh/roadshow/glossary/).
8. Lancashire Record Office, Hornby Catholic Mission Papers (St Mary's Church), Correspondence Ref: RCHY 3/7 1775.
9. Harriet Guest, "A double lustre: Femininity and sociable commerce 1730–1760", *Eighteenth Century Studies*, Volume 23, Number 4 (Summer 1990): 481.
10. B.R. Haydon, ed., Simon Brett. *The Faber Book of Diaries* (London & Boston: Faber & Faber, 1987) 320.
11. For example, see the Garrard Account Books, microfilm in National Art Library, customer ledgers, 1735–1819.
12. Ralph S. Walker, ed., *A Selection of Thomas Twining's Letters 1734–1804. The Record of a Tranquil Life,* Vol 1 (1791; Lampeter: Edwin Mellen Press, 1991) 382.

12

Archaeological Conservation: Scientific Practice or Social Process?

Elizabeth Pye

Introduction

Archaeological conservation is concerned with sites, structures, and associated artefacts that are the focus of archaeological study. This chapter concentrates on conservation applied to objects discovered through excavation, and aims to examine the extent to which archaeological conservators are bound by the same general principles as other conservators. The constraints of the context (conservators often work in the field) and the condition of the material (frequently highly deteriorated and sometimes unrecognizable) may limit the ways in which accepted conservation principles can be applied.

Principles represent the agreed philosophy of a profession, but achieving agreement and definition of principles takes time, so existing codes may lag behind current practice and evolving thought. The philosophy typical of archaeological conservation has a scientific focus reflecting the way in which this branch of conservation developed in the late nineteenth and the twentieth century as a scientific practice (linked to a scientific approach to archaeology).[1,2] However, since the late twentieth century there has been a shift in archaeology, and more recently in conservation, towards a more inclusive social approach. A second aim of this chapter is to examine the extent to which archaeological conservation is scientific practice or social process.

The archaeological context

Archaeology aims to develop hypotheses about activities and life in the past through interpretation of material remains. Characteristic of archaeology is a methodology based on excavation and analysis of stratified deposits and the evidence (including objects) encapsulated in the strata.[3] Counterbalancing this analytical practice, and increasingly considered important, are the social aspects of archaeology, involving all those with an interest in archaeological activities and the material remains of the past. 'Public archaeology' embraces actions at international and national level in the protection and study of heritage. It also embraces more informal interests associated with tourism, local societies, personal collections, and activities such as metal detecting; these are often exploited and encouraged by the media, particularly television.[4] Whereas 50 years ago archaeology was a relatively narrow specialist interest, now many people, with sometimes-conflicting views, are involved – including politicians, religious groups or local communities.[5] The acceptance of multiple views is also linked to the recognition now given to intangible heritage, such as language and drama.

There is a general presumption that archaeological remains are best safeguarded *in situ*, and it is often argued that excavation should take place only when a site is threatened (although research interests may justify excavation of unthreatened sites). However, it is increasingly difficult to conserve sites in this way as they are threatened by land changes caused by building development, drainage, or industrial farming methods. The trade in illicit antiquities, exacerbated by ease of communication through the Internet, and fed by demands of collectors, has also become a major problem.[6] Increased personal mobility has boosted tourism, which damages sites through over-visiting. Climate change is likely to damage buried sites through rise or fall in the water table, or through erosion caused by storms and flood. As conservators we need to be aware of the role we can play in limiting this damage – for example, by striving to limit carbon emissions or by refusing to work on illicit material.

Archaeological objects

Excavated objects often display more extreme deterioration than other types of objects. Some soil conditions are so aggressive that little material survives, or what does survive is in a very poor state; there is often a bias towards survival of inorganic materials (such as ceramics, metals) while organics (including wood, textiles) leave little or no trace.[7] Only in particular contexts is there excellent preservation of the whole range of materials, as in waterlogged or under-sea conditions (in the case of artefacts retrieved from wells or ship-wrecks)[8] or as in the very dry conditions of Egypt (objects displayed in the Cairo Museum show staggering preservation). In many contexts, the generally poor survival of organic materials means that even quite tiny and apparently unattractive traces of wood or textile can be important as sources of information.

The relation of an object to its context, and to other objects in an assemblage, is crucial in developing ideas about types and sequences of activity, and in dating a site, so individual objects can be likened to pieces in a jig-saw – significant as part of a whole. Objects provide the possibility of reconstructing technological activities, and of answering questions such as: How was this made? But this is not simply a dry scientific process of technical analysis and typological dating – it also contributes to developing ideas about past people's motivations and behaviour. We can begin to consider questions such as: What was it like to be alive then?

An excavated object is likely to be far from its 'as made' state. Its significance as archaeological evidence is intimately linked to its material character embracing not only the original substance but also the indications of all the material changes that have happened as a result of manufacture, use, discard, burial and discovery. We sometimes refer to this totality as the 'physical integrity' of an object, and are cautious about any action that might affect it. Paradoxically, however, investigation and conservation processes may involve either removal or addition of material, and thus modification of the object's form or composition.[9,10] Changes that happen after excavation, such as renewed corrosion, may not be considered aspects of this integrity, but increasingly we consider earlier conservation treatments to be significant aspects of the history of the object.

For all these reasons, careful judgement is needed to decide where the archaeological significance of an object may lie, particularly because there is probably more information in some objects than we have yet been able to tap. For example, in the last 20 years or so, it has become possible to identify traces of foodstuffs, thereby indicating what ceramic vessels contained, and to examine DNA in bone, thus detecting relationships between buried individuals (and future innovations will presumably enable us to learn even more).[11] It is an important principle that the aims, processes and results of investigation and conservation are fully documented so that future re-investigation or re-treatment can take account of earlier work.

Archaeological conservation

Archaeological conservators work in the field where their task is to limit the deterioration that is often activated by excavation. We also work in museums on recently excavated material, or on existing collections, where we normally focus on investigation, elucidation and treatments aimed at preventing further change. Apart from spectacular discoveries, comparatively few objects from current excavations are considered suitable for public display; so restoring objects is a relatively minor aspect of many archaeological conservators' responsibilities. By contrast, conservators working on major archaeological collections, such as those of the British Museum, are routinely involved in preparing significant material for exhibition. However, in this as in other fields of conservation, there can be overlap, both in intention and effect, of different conservation techniques. Reconstructing a damaged object may stabilize it, but also clarify its form for specialist study, as well as making it more visually accessible to visitors.[12,13,14,15]

Very large quantities of artefacts (frequently highly fragmentary) can arise from excavation, so most objects are studied as populations rather than individuals. This has several effects: we make a preliminary selection (we often use X-radiography to select particular metal artefacts for detailed investigation), or we treat large quantities of material in batches (regularly the case for waterlogged leather fragments).[16] Although sometimes argued by conservators in other fields that every object should receive the same level of attention, it is simply not feasible for us to apply a uniform standard of conservation treatment to the bulk of archaeological material.

Preventive conservation in the field

An important aspect of archaeological conservation is the preventive measures undertaken in the field. If remains are left undisturbed in the ground, equilibrium is reached and deterioration slows or even ceases. The process of excavation upsets this equilibrium and puts objects at risk: it may induce rapid change through exposure to oxygen, sunlight, changes in humidity, and to loss of the physical support provided by surrounding soil. Exposure is particularly risky for waterlogged organics – loss of water through evaporation may result in fragmentation of surface detail and structural distortion or collapse – or for highly desiccated objects, which may simply crumble.[17]

Our objective is to minimize the shock of excavation. As far as possible we replicate burial conditions: so waterlogged material is kept wet, and fragile objects are given alternative physical support to take the place of the surrounding soil. Although we use preventive measures as far we can, some fragile material may need a rather more interventive approach, such as application of adhesive and temporary facing in order to provide support during excavation.[18,19] We accept that the constraints of working in the field, often with limited time and resources, necessitate adaptability, especially in an 'emergency,' when significant objects are discovered unexpectedly. So here, too, we may not always be able to use the 'best' approaches and materials.

Artefact investigation: the conservator's responsibility

The importance of investigation is a consequence of archaeology's focus on material remains as evidence of past activities. Excavated archaeological objects can be so deteriorated that their investigation is likened to forensic detective work, and may involve X-radiography to elucidate interior structure,[20] cleaning to reveal surface detail,[21] or analysis of accretions. Investigation has become our responsibility because discovery of evidence must go hand-in-hand with ensuring that the evidence is not damaged or lost, or that its elucidation does not distort other actual or potential information.

Investigation plays a crucial role in aiding interpretation of the object for the specialist. Preliminary cleaning can be compared to archaeological excavation and we often refer to it as 'investigative cleaning' (as in cleaning a coin in order to identify and date it). Being destructive it can be a risky and potentially controversial activity, so we must weigh carefully both the benefits and the risks. We must assess the nature of accretions: whether they are extraneous – such as adhering soil – or products of deterioration of the object itself, and we must evaluate the effects of deterioration. In some materials (for example, glass) deterioration changes surface appearance but not volume, so we retain the deteriorated surface, but in other materials (for example, iron) deterioration can result in voluminous corrosion, which we generally remove as it obscures the object. However, metal corrosion may not only obscure surfaces but also contain information, such as tiny traces of mineralized wood or textiles; furthermore, surface detail may lie *within* the corrosion layers rather than beneath them, so we must be vigilant for this kind of information. We also need to discriminate between foreign accretions and original features: a dark material in engraving on a bone object may be 'dirt' or black pigment; a chalky-looking deposit on a decorated copper alloy object might be degraded enamel. Deciding exactly what to remove and where cleaning should stop requires careful judgement; so we may clean objects only partially in order to leave material for future re-investigation.[22]

We use other forms of investigation in order to understand composition and techniques of manufacture 'recorded' only in the objects themselves. Results of examination can be indicative of innovation and evolution in technologies (for example, the makers' ability to harness fire, or to achieve precise control of temperature) and may contribute to wider archaeological theories about, say, contact between groups. For these reasons we must be familiar with a range of analytical techniques, from simple chemical spot tests to instrumental analysis such as scanning electron microscopy or X-ray diffraction.

Remedial conservation

When using remedial treatments we aim to stabilize existing physical damage and reduce active deterioration. Broadly, we divide treatments into those that involve adding material to enhance stability, and those that involve removing materials that are sources of active deterioration. Examples of added materials include modern synthetic polymers used as adhesives or consolidants, or corrosion inhibitors applied to metals. In practice they are considered to be more or less permanent additions, as attempts to remove them would put fragile archaeological objects at risk; so reversibility has been an ideal, but never an actuality. We aim to choose materials that we can be reasonably confident will not damage the object and distort the known or potential information it carries, so we must have a good knowledge of the properties and effects of these additives, particularly as some have proved to be unreliable or hazardous.

We use many treatments to remove harmful substances, including soaking in order to dissolve soluble salts that have penetrated the pores of a ceramic, or disinfection of an object affected by fungal spores. An even greater degree of intervention involves changing the material of the object itself. We may use this when degradation has altered the character of the object so profoundly that little coherent form is left. For this reason, electrolytic reduction has sometimes been used to convert lead corrosion to a more homogeneous metallic state on objects that have suffered in poor storage. This approach does not conserve the object as found, and alters evidence of original technology, so is controversial. In each situation we must weigh up the effects of treatments on the perceived archaeological integrity of the objects and use detailed documentation to record both process and effects. We use all treatments as cautiously as possible but there may be situations where unstable excavated artefacts will not survive without an extent of remedial action normally considered over-interventive in other fields of conservation.

Preventive conservation of stored material

A major problem is presented by the large amounts of material being excavated and what is seen, in some parts of the world, as the consequent crisis in the size of the stored archive.[23] This situation has increased our responsibilities since the objective of the archive is that material should be preserved for future study. We aim to create favourable conditions for long-term preventive conservation through providing effective packaging and establishing and maintaining a suitable environment. However, the large quantities of material make it difficult to monitor the condition of objects regularly, so the state of stored material frequently has to rely on the passive effects of good packaging and environmental conditions.

Interpretation through restoration

Relatively few excavated objects are exhibited, thus many remain accessible to specialists alone. We use restoration techniques only where an object is particularly significant or can be used in display to communicate information about a site, period, or activity. Techniques are intended to reinstate something of the original appearance, and normally embrace cleaning, reconstruction from fragments, and completing missing areas or features.

We hold that restoration should not alter materials, or conceal the effects of use, discard and burial. As far as possible, fills or reconstructed features are designed to be readily removable and are toned to the general colour of the fabric rather than matched precisely, the principle being that viewers should be able to distinguish original from restoration.[24,25] Furthermore, we do not reconstruct or reshape objects if the damage relates to their original use or to their deposition (such as apparent ritual breakage of weapons), or reflects significant events relating to their discovery. In any case, reshaping can be controversial because of the risk of losing technical evidence.[26]

Re-conservation and re-interpretation

The archaeological archive includes long-standing collections in museums. Many of these objects were discovered or acquired during the last two or three centuries, and restored according to the practices of the time. Our work can require deconstruction of old restorations that may be failing or causing damage, or re-investigation to understand the objects better. There is a potential conflict here – interest is increasing in the history of conservation, but removal of old restorations removes the evidence of earlier practices.

Conservation as social practice

The scientific, material-focused approach to archaeological conservation has been established for well over half a century, but recognition of wider public interest in the past has exposed us to new views and pressures. It is now acknowledged that excavated objects can have many different intangible meanings (such as personal, political, aesthetic or religious) for people today. What may be a piece of evidence to an archaeologist, or conservator, may have deep spiritual meaning for a member of a descendant community. A particularly telling example is that of human remains – seen on the one hand as specimens stored for scientific analysis, and on the other as ancestors who have been exhumed and denied the right of burial.[27,28]

We must consider these differing values carefully when working towards conservation decisions. Whereas 20 years ago, scientific factors would have governed our thinking, now, to the emphasis on material or physical integrity, we must add consideration of the values that compose an object's intangible cultural significance.[29] Furthermore, we must reach a balance between values that sometimes conflict. Should human remains be investigated and conserved, or returned to the relevant community for reburial? In the USA, the Native American Graves Protection Act (NAGPRA)[30] has enforced the return of human remains together with their accompanying grave goods, and much conservation work has focused on preparing material for return. Should a feature that has 'always been there' be excavated, conserved and studied, or simply left alone where it 'belongs'? In one case, local people in a Scottish community were in favour of reburying the long-lost base of a famous stone cross-slab, despite the upper part being a valued exhibit in the Museum of Scotland.[31]

These examples reflect not only the strength of personal feelings, but also a view that objects have lives, and a fear that archaeologists' or conservators' interventions may rupture the natural course of these lives. In fact, as archaeological conservators we have long understood that objects are not static, that both material and meaning can be changed by events such as excavation, and that it is possible to manage material change, but seldom to eliminate it. Furthermore, we acknowledge that intangible significance may be linked to material change (as in the value given to patina). We now also recognize that conservation practice itself, far from being 'neutral,' contributes to the unfolding life of an object by instigating material change, or by giving preference to a particular meaning. We also accept that views of significance are not firmly anchored in the materials of an object but may shift with new audiences and changing interests.

Public questioning of conservation practice also indicates a suspicion of experts who, certainly until recently, have been seen as exclusive. However, archaeology is becoming increasingly inclusive, as seen in such activities as community excavations. Some archaeological archives are now open to visitors, and rather than guarding their professional expertise, archaeologists and conservators welcome the involvement of volunteers, and work with groups such as metal detector users, identifying their finds and advising on their care.[32,33] We aim to display conservation activities openly rather than screening them from public view, and visitors' questions are welcomed. Communication with the public has become as important as communicating with fellow specialists; so although we have been accustomed to using formal scientific reporting and specialist terminology, we are now beginning to use everyday language and to 'tell the story' of a conservation project.

Conclusion: principles of archaeological conservation

What can be said to typify archaeological conservation? It has been profoundly affected by an emphasis on material evidence and investigation, and on the desire to re-investigate in the future. The large quantities of material coming from some excavations lead to the necessity of selection for both investigation and treatment, or alternatively to the need for bulk conservation treatments. An understanding of archaeological context is essential in order to focus field treatment and subsequent investigation appropriately (and to discriminate against illicitly obtained material). Although the concept of minimum intervention is important, the nature of material and context may require quite interventive approaches. At the same time, public involvement and new audiences are shifting philosophy and practice towards a greater emphasis on intangible meanings of objects, and provision of wider physical and intellectual access.

For us, conservation increasingly involves negotiating a balance between apparently or actually opposing positions – between protection of and access to objects, between preservation for future use and use now, and particularly between the needs of science and the interests and beliefs of people. In some situations opposing positions lead to controversy and conflict, and human remains are a poignant example. In 1989, the Vermillion Accord was drawn up in an attempt to encourage indigenous peoples and scientists to respect each other's views. Paragraph 5 states that 'agreement . . . shall be reached by negotiation on the basis of mutual respect for the legitimate concerns of communities . . . as well as the legitimate concerns of science and education.'[34] This accord is a useful expression of the need for respect and negotiation in potential conflicts between the scientific and social approaches to conservation. The requirement for respect and negotiation may be increasingly important in the future as climate change confronts us with a conflict between the urgent needs of human populations and our desire to preserve our archaeological heritage.

Acknowledgements

I am most grateful to Nick Balaam, Janey Cronyn and Helen Ganiaris, as well as to the editors of this volume, for their helpful comments and suggestions.

Notes

1. H. Plenderleith, "A history of conservation," *Studies in Conservation* 43 (1998): 129–143.
2. M. Gilberg, "Friedrich Rathgen: the father of modern archaeological conservation," *Journal of the American Institute of Conservation* 26 (1987): 105–120.
3. C. Renfrew and P. Bahn, *Archaeology: Theories, Methods and Practice* (London: Thames & Hudson, 2004 edn).
4. N. Merriman, *Public Archaeology* (London: Routledge, 2004).
5. B. Bender, *Stonehenge: Making Space* (Oxford: Berg, 1998).
6. N. Brodie, ed., *Archaeology, Cultural Heritage and the Antiquities Trade* (Gainesville: University Press of Florida, 2006).
7. J. Cronyn, *The Elements of Archaeological Conservation* (London: Routledge, 1990).
8. C. Pearson, *Conservation of Marine Archaeological Objects* (London: Butterworths, 1987).
9. M. Berducou, "Introduction to archaeological conservation," *Historical and Philosophical Issues in the Conservation of Cultural Heritage*, eds., Nicholas Stanley-Price, *et al.* (Los Angeles: The Getty Conservation Institute, 1996) 248–259.
10. E. Pye, *Caring for the Past: Issues in Conservation for Archaeology and Museums* (London: James and James 2001).
11. C. Caple, *Objects: Reluctant Witnesses to the Past* (Abingdon: Routledge, 2006).
12. Cronyn, 1990.
13. M. Berducou, ed., *La Conservation en Archéologie: Méthodes et Pratique de la Conservation-Restauration des Vestiges Archéologiques* (Paris: Masson, 1990).
14. C. Caple, *Conservation Skills: Judgement, Method and Decision-making* (London: Routledge, 2000).
15. Pye, 2001.

16. B. Wills, ed., *Leather Wet and Dry: Current Treatments in the Conservation of Waterlogged and Dessicated Archaeological Leather* (London: Archetype for the Archaeological Leather Group, 2001).
17. R. Brunning, *Waterlogged Wood: Guidelines on the Recording, Sampling, Conservation and Curation of Waterlogged Wood* (London: English Heritage, 1996).
18. D. Watkinson and V. Neal, *First Aid for Finds* (London: Rescue – The British Archaeological Trust and Archaeology Section of the United Kingdom Institute for Conservation, with the Museum of London, 1998 edn).
19. C. Sease, *A Conservation Manual for the Field Archaeologist* (Los Angeles: Institute of Archaeology UCLA, 1994 edn).
20. F. Shearman and S. Dove, "Applications of radiography in conservation," *Radiography of Cultural Material*, ed. J. Lang and A. Middleton (London: British Museum, 2005) 136–154.
21. Caple, 2000.
22. Pye, 2001.
23. N. Merriman and H. Swain, "Archaeological archives: serving the public interest?" *European Journal of Archaeology*, Volume 2 (1999) 249–267.
24. S. Koob, "Detachable plaster restorations for archaeological ceramics," *Recent Advances in the Conservation and Analysis of Artifacts*, ed., J.W. Black (London: Summer Schools Press, 1987) 63–66.
25. S. Watkins and R. Scott, "Timeless problems: reflections on the conservation of archaeological ceramics," *Past Practice, Future Prospects: British Museum Occasional Paper 145*, eds., A. Oddy and S. Smith (London: The British Museum, 2001) 195–199.
26. Caple, 2000.
27. E. Pye, "Caring for human remains: a developing concern," *Past Practice, Future Prospects: British Museum Occasional Paper 145*, eds., A. Oddy and S. Smith (London: The British Museum, 2001) 171–176.
28. C. Fforde, *Collecting the Dead: Archaeology and the Reburial Issue* (London: Duckworth, 2004).
29. M. Clavir, *Preserving What is Valued, Museums, Conservation and First Nations* (Vancouver, Toronto: UBC Press, 2002).
30. NAGPRA *Native American Graves Protection and Repatriation Act* Public Law 101-601 (U.S. Congress 1990).
31. S. Jones, "'They made it a living thing didn't they…'. The growth of things and the fossilization of heritage," *A Future for Archaeology*, eds., R. Layton, S. Shennan, and P. Stone (London: UCL Press, 2006) 107–126.
32. H. Ganiaris and L. Goodman, "Piloting a new volunteer programme," *Icon News* March 2007: 46–47.
33. R. Hobbs, C. Honeycomb and S. Watkins, *Guide to Conservation for Metal Detectorists* (Stroud: Tempus, 2002).
34. Vermillion Accord 1989 (available at www.worldarchaeologicalcongress.org).

13

Conservation and Cultural Significance

Miriam Clavir

Franz Boas spent a decade of intense involvement with the American Museum of Natural History and then resigned in 1905, convinced that it was impossible to adequately represent cultural meaning on so slim a basis as physical objects.[1]

Conservation professionals are directed by their codes of ethics to preserve the cultural significance of material heritage under their care. Likewise, it is due to this significance that the material is being preserved. Tangible and intangible qualities are included as meaningful: for example, 'aesthetic, artistic, documentary, environmental, historic, scientific, social, or spiritual values.'[2]

One problem, however, is that '[t]he field of conservation deals with the physical aspect of cultural property,'[3] and 'cultural significance' is a social construct. Conservation professionals are supposed to 'refrain from making statements based solely on opinions not rooted in physical evidence.'[4] Conservators by their very training and principles attempt to avoid subjective interpretation, yet cultural significance is society's interpretation of what is important.

How then is a conservation professional expected to know, in looking at an object, which qualities are to be preserved because they are culturally significant?[5]

'Declarations of age, origin, or authenticity should be made only when based on sound evidence.'[6] But even if there is strong physical evidence of these attributes, how does the conservator know they are meaningful?

Richard Handler, for instance, provides examples showing that even proven facts such as date of attribution and the artist's name are not necessarily significant as they have no meaning unless interpreted.[7] A work thought to be by 'X' that is shown to be 'by the school of X' is the same physical work, but it has been interpreted and reinterpreted to give it more or less meaning and significance. Even if the signature is authentic and the date known, Handler distinguishes between facts, and meaningful facts.

> What is at issue here is not the possibility of 'facts' or factual knowledge. I do not doubt that it is possible to identify the individual person who created

a particular object, as well as the date of its creation; these are facts that are sometimes possible to know about an object. My point, however, is that such facts have no meaning outside an interpretive frame of reference. Therefore, to the degree that such facts lead us to attach value to an object, that value is not based solely upon something intrinsic to the object but upon the place the fact assumes in a broader theory of value . . . The fact of the object's date is not intrinsic to the object but is part of a meaningful narrative that we construct in interaction with that object.[8]

He concludes, 'The life of objects is social, not objective.'[9]

Whether one accepts his arguments or not, conservators are charged in most codes of ethics with preserving for the future of society, usually taken to mean 'society as a whole.' What if you are a member of a minority group that disputes, for a given object, what is considered important, iconic, significant?[10] What is celebrated by the larger society may have a vastly different meaning to minority communities. It is only relatively recently in Canada, for example, that official national historic sites have been designated and/or interpreted from an aboriginal point of view, in addition to the earlier-registered trading and military forts of Canadian history.

- Who gets to choose what is culturally significant?
- Have objects safely crossed into culturally significance once they have been catalogued into a museum's collection? Or is this collecting basically a reflection of the museum having a curator whose life work was in that area?
- Are sites culturally significant because they have been deemed so by government ministries or organizations? Or does designation freeze meaning, with relevance potentially diminishing generation by increasingly globalized generation?
- When contemporary media inform us about artistic works, the marker reported is often predicated on sales. Do high auction prices validate a work of art as significant? Or high museum entrance receipts for an exhibition? Perhaps success today is simply having been noticed by the media. Is cultural success the same as cultural significance?

In all these cases, tongue-in-cheek or not, cultural values are deeply felt, interpreted, superimposed and subsequently read as the truth; yes, this work of art, object or site is important. The designation of 'importance' may be the result of, and result in, broad public or institutional recognition. Significance may also come from local community or smaller-group collective recognition, and it may come from individual experience and life-histories. Sarah Harding states:

> . . . the things and places we identify as 'cultural property' . . . are the products of and reflect our collective experiences in their creation, in their formal dedication, and in the on-going reinscription of their meaning. . . . But the significance

of much (if not most) cultural property and heritage originates not in the public realm but in personal experiences, everyday life, and local contexts.[11]

Harding point outs that cultural property becomes meaningful through personal as well as collective experience. This is not only to state, for example, that family memorabilia are highly significant to those individuals. Harding is also saying that, for instance, a landmark building or work of art is significant not only because a community has deemed it thus, but becomes personally significant for every visitor or viewer each time they experience it. 'The significance of cultural property is the product of multiple interactions.'[12]

This point concerns conservators. Preventive conservation arrests the deterioration of collections, but by its nature and goals often limits these 'multiple interactions.' Is it not hypocritical for conservators to state their desire to preserve the culturally significant aspects of material heritage when their professional field usually gives them neither in-depth participation in contemporary debates surrounding cultural significance, identity, loci of memory, tradition/authenticity, nor visitor access as a required consideration – unlike libraries, for instance – where hands-on access even to most Special Collections is a necessary principle in the collections' physical preservation?

Yet, is this the realm where conservation principles are subtly changing? Compromises are increasingly being recognized as acceptable professional practice. David Leigh states,

> More recently, however, those simple ideas and principles on which many of us [conservation professionals] here today cut our teeth – for instance reversibility, authenticity, integrity, true nature, original surface, cleaning – have begun to melt away, or at least go fuzzy at the edges. We have seen them subjected to scrutiny and in daily need of refining.[13]

Most conservation codes of ethics have included a statement speaking to the necessary balance between a society's need to use cultural property and the preservation of it. Many, however, like the AIC Code of Ethics, also state, 'While recognizing the right of society to make appropriate and respectful use of cultural property, the conservation professional shall serve as an advocate for the preservation of cultural property.'[14] The National Trust in the UK, though, places the emphasis slightly differently, on conservation as 'the careful management of change.'[15] Sarah Staniforth points out that in the UK, this current definition of conservation from the National Trust represents an attitudinal change from its 1907 Act: 'promoting the permanent preservation for the benefit of the nation . . .'[16]

These examples of shifts in thinking are not the only changes conservators have to consider. As a social construct, 'cultural significance' can change as well, and if a museum object becomes more or less meaningful, this can have an impact on conservation. For example, if a particular bird is now extinct, do remaining natural history specimens deserve upgraded conservation measures? A utilitarian canoe became a cultural icon to a Canadian aboriginal community because it was the last one left.[17]

Cultural significance is tied not only to the time period that created its original meaning, but to today, and the future. This is acknowledged for pieces believed to have continuing sacred or culturally sensitive attributes, and is an area in which conservators today recognize the appropriateness of cultural protocols in conjunction with conservation. Awareness may be more difficult when the cultural significance being preserved today is not tied to the object's original purpose or meaning, such as the example given above of the canoe. Do conservation's risk assessment procedures allow, in a cost-effective way, for changes in or additions to cultural significance over time?

Significance is sometimes not apparent even when considering only the object's original meaning. Many aboriginal baskets in museum collections were made for sale, while some were made for indigenous daily use or ritual use. Sacred/sensitive baskets may be recognized by ceremonial practitioners. Some baskets, however, may not be distinguishable from 'ordinary' baskets unless evidence of associations, such as the placenta of a newborn, are still there.

Sometimes changes are mainly to the museum staff's expectations of what is significant. In the following example, the cultural significance of the object is unanticipated, and the significance of conceptual frameworks that govern museum practice is challenged.

> When the [Portland Museum of Art] made plans to reinstall and reinterpret the [Rasmussen] collection in the late 1980s, they decided to involve . . . a dozen prominent Tlingit elders, representing clans with specific relationships to the objects. . . . [O]bjects were brought out and elders were asked to interpret and speak about them. Clifford describes how he and the curatorial staff, focusing on the objects, waited expectantly for some sort of detailed explication about how each object functioned, who made it, and what powers it had within Tlingit society. Instead, he reports, the objects acted as memory aids for the telling of elaborate stories and the singing of many songs. As these stories and songs were performed, they took on additional meanings. An octopus headdress, for example, evoked narratives about a giant octopus that once blocked a bay preventing salmon from reaching inland rivers. By the end of the story, the octopus had become the State and Federal agencies regulating the right of Tlingits to take salmon; so that what started out as a traditional story took on precise political meanings in terms of contemporary struggles. 'And in some sense the physical objects, at least as I saw it, were left at the margin. What really took center stage were the stories and songs.'[18]

Conservators and other museum professionals today recognize the significance of intangible cultural heritage, but it is harder to acknowledge that the objects in collections may have little or no importance. Julie Cruikshank discusses examples of the minimal meanings to their makers of some aboriginal objects that have found their way into museums, at least when these objects are separated from other aspects of their lives. For example, in comparing the world-renowned, highly visible, and symbolic material culture of the First Nations of the Northwest Coast of British Columbia (for example, totem poles, masks), and those groups living in the Interior of the province and western subarctic, she says of the latter:

> Yet for people whose successful harvesting of resources depended on strategies of mobility, priorities differed. They were more inclined to carry essential ideas in intricately woven narratives than in cumbersome material objects. More important than the physical object was the ability to recreate a snare or a container or a shelter when and where one was needed.[19]

Cruikshank relates how several carvings in the MacBride Museum in the Yukon, made by a Tlingit/Tagish elder, Kitty Smith, were products created as she told traditional stories. The narratives had personal significance as well as cultural significance:

> During her life, she would not isolate discussions of these experiences [personal experiences related to family economic dislocation, illness and death] from her accounts of how the world began, how humans and animals came to coexist and how she carved these stories in wood. Such things, she would say, simply cannot be understood separately. Her carvings actively contest not only the categories used to display culture in museums but also the idea that 'traditional' culture exists in any static sense. . . . The stories Mrs. Smith told to describe her work refer to the act of creation rather than to the finished object: keeping them would have been highly impractical for anyone with a lifestyle as mobile as hers. Her carvings were not discrete 'things' but one part of a tradition she used to engage with the world around her.[20]

To their maker, the carvings had significance both personal and larger, during their creation but not afterwards. For her daughter, seeing them over fifty years later in a museum, they invoked memories of what her mother could do, but when asked about her mother's carving techniques and materials (typical questions a conservator might ask), she began the same stories her mother had told underlying each carving.[21]

Objects are part of an on-going dialectic that reflects, interprets and shapes and reshapes lived experience and understanding. Significance is formed from the 'multiple interactions' referred to earlier, including the acts of creating or using objects.

To summarize so far, meaning is not fixed, nor is it necessarily compatible with ideas in conservation. 'In the Zuni world-view meaning and significance may change by time of year or by virtue of ceremony.'[22] Nancy Marie Mithlo quotes Edmund J. Ladd, a Zuni museum ethnologist and curator:

> We believe that things that are put in museums will eventually eat themselves up 'EEWETONAWAH'. In other words, they will completely disintegrate and do their own thing anyway no matter what the museum does to preserve it. We are saying to the museum 'Keep them because we know better.' We say to the museums 'If you return [sacred objects] we will curate them according to our traditions. According to our traditions, they have to be put into the ground and destroyed.' Curate simply means to take care of. We take care of it the way it is supposed to be taken care of. Preservation is not a part of Zuni culture. Preservation is completely opposite of our concept of deterioration and disintegration as a means of refurbishing and re-entering into [the] afterworld.[23]

Mithlo comments that the Zuni well understand the differences between museum value systems and their own, and believe theirs will prevail. 'The pueblo has even appropriated the use of the term "curate" and altered its meaning to fit their social reality.'[24] Likewise the museum meaning of 'preservation' is turned on its head for sacred/sensitive pieces.

Leigh, quoted earlier, referred to shifts in ideas in conservation. To give an additional example, many conservation professionals have enlarged their perspective from a strictly scientific definition of deterioration – for instance:

> [A] degradative adjustment of materials to the conditions prevailing in their immediate environment: existing chemical compounds are converted in the process of deterioration to compounds of greater stability vis-à-vis their surroundings.[25]

to include a social definition of deterioration, such as:

> 'We describe deterioration as those changes that we regard [as] undesirable.'[26]

This shift acknowledges that for some objects, such as the sacred/sensitive pieces Ladd is concerned with, or certain patinas, deterioration can be described as desirable.[27]

At the UBC Museum of Anthropology in Vancouver, Canada, staff use preventive conservation measures to preserve a large weaving from the Salish community of Musqueam that is on permanent exhibition. The weavers themselves have stated that they do not mind if the weaving fades or its weight pulls it down unevenly. When that happens, it is time for the Museum to commission a new weaving to be made, thus actively helping to preserve the heritage of weaving through the generations.[28]

In ethnographic and archaeological conservation at least, conservators are being asked to understand not only which 'special qualities' of an object or site should be 'protected, enhanced, understood and enjoyed by present and future generations'[29] but also, to consider that cultural significance may lie in the physical object not being preserved at all, even when it constitutes artistic achievement or the only remaining 'real evidence,' the actual witness of and from the past.

The question of 'real evidence,' authenticity, and tradition are discussions that have an impact on conservation and its goals. Concepts, such as preserving the 'integrity' or 'true nature' of an object, have already had an evolving history in conservation. Pertinent to the discussion here on cultural significance is the argument that conservation inevitably changes the objects being worked on; if their 'integrity' or 'authenticity' is being changed by conservation practice, is their meaning being changed?

> Pearce (1990) discusses the complex relationship between (1) the appearance of archaeological objects before and after treatment and (2) archaeological information. She says that 'the object as it emerges from the ground is an encapsulation of its history up to that moment; but the unravelling of that history by the modern investigative techniques of the conservator inevitably involves the destruction of evidence as much as the preservation of a version of the artefact'.[30]

> Conservators may believe that they are revealing the 'true nature' of the object; however, Pearce believes that what they are actually revealing is a version of the object – one in which the irreversible processes inherent in the excavation, cleaning, and consolidation of archaeological materials preclude the possibility of verifying information that may have been present pre-excavation or pre-treatment.[31]

Debates about what constitutes authenticity or tradition belong to a longer, separate discussion, and the reader is referred to other publications.[32] These concepts, however, can be important in conservation decisions; as one example, the National Trust argues cogently why it keeps the properties under its care in different states of restoration.[33]

Conclusion

The reader is forgiven for believing we have arrived at a low point in this conversation where, depending on one's point of view and the piece in question, she or he reaches the conclusion that objects in museums may not be culturally significant, the act of conserving them at best has a good chance of altering them and at the worst (for conservators) is not important anyway, and the training conservation professionals receive is inadequate to understand this.

In practice, the situation is rarely this extreme, nor should differences in perspective build to such an impasse. (For successful consultations between conservators and indigenous groups, for example, see Johnson, Heald, McHugh, Brown and Kaminitz.)[34]

Ruth Phillips, an art historian, writes:

> I think the answer lies at least in part in realizing that in the past, we thought of an object as a sort of a self-contained package of information, from which data could be retrieved. Today objects are viewed as material embodiments of social and cultural perspectives and relationships that are increasingly multi-vocal. Objects are made by people and used by people for cultural purposes. Objects have dynamic cultural biographies that do not end but only change when they enter museum collections.[35]

Conservation professionals began as experts in the scientific and technical make-up, deterioration, stabilization and often restoration of works of art and artefacts, and have expanded their knowledge to include a contemporary awareness of social constructs affecting collections. 'Multiple interactions' within and between understanding the physical object and the social object produce a rich interface where today multi-dimensional, multi-collegial approaches are used to advance conservation decisions. To take just one example from the papers published at the Munich 2006 IIC Congress, *The Object in Context: Crossing Conservation Boundaries*:

> [This paper] addresses how some conservation boundaries were crossed, in order to contribute to a better understanding of life during the First World War, and discusses how material culture is valued differently in different contexts (and how this will influence conservation decisions). It concludes that neither object meaning nor conservation decisions can be viewed objectively and that conservation has to be viewed as a social process governed by economic, political, religious, social and cultural dynamics, rather than a primarily technical process.[36]

When conservation is practiced in museums, it participates in the values of the museum as an institution, what Michael Ames calls 'The Idea of the Museum.'[37]

> This Idea of the Museum, more correctly a complex of ideas, encodes two fundamental principles that have far-reaching implications for the museum

movement: (a) that collections are vital to the understanding of heritage, thus they should form the focus of museum work, and (b) this work is a moral good every community should respect and desire. These principles represent a notion of culture that infuses technical procedures with moral imperatives.[38]

Conservation is a complex of values, knowledge, skills and processes guided by the moral imperatives of its codes of ethics. Conflicts have arisen when moral imperatives are accompanied by 'high moral ground' attitudes, or when one's own 'moral imperative' is imposed on others, especially those outside of the conservation or museum system.

Conservation is also the child of, situated in, and continually influenced by its surrounding social environment. In many cases this is the same powerful environment that has defined cultural significance for collective cultural property in western societies. Broadly shared cultural meanings, however, are not the only definition of cultural significance, just as conservation is not a universal definition of 'best practice' preservation.

> [B]oth Indigenous knowledge and Western knowledge systems can be interpreted as subjective enterprises with restricted codes. Museum mandates to collect and preserve are not universal standards but particular norms associated with specific embedded social histories.[39]

A question posed at the beginning of this chapter, 'how is a conservation professional expected to know which qualities are to be preserved because they are culturally significant?' cannot be given a template-like answer. At the same time, evidence of the social construction of conservation and museums should never be taken as negating the worth of conservation and museums. 'Museums serve useful purposes. . . . They do not serve all purposes, however.'[40]

Shifts in conservation thinking resulted from examination and re-examination of the norms and principles of the profession as well as from scientific evidence and daily practice. Every conservation code of ethics discusses 'analysis' or 'examination' as a fundamental part of conservation. Broadening these words to include reflection on social and cultural concepts as debated in other disciplines enables conservators to better participate in decisions in the preservation of 'cultural significance.'

Notes

1. Julie Cruikshank, "Imperfect translations: Rethinking objects of ethnographic collection," *Museum Anthropology*, Volume 19, Number 1 (1995): 25.
2. European Confederation of Conservator-Restorers' Organizations (ECCO), Preamble, *Professional Guidelines*, adopted by its General Assembly Brussels 1 March 2002.

3. American Institute for Conservation of Historic and Artistic Works (AIC), "Commentary 18: Interpretation," *Code of Ethics*, revised 1994.
4. American Institute for Conservation of Historic and Artistic Works (AIC) 'Recommendation' for AIC "Commentary 18: Interpretation," *Code of Ethics*, revised 1994.
5. In this text the word 'object' is used as a generalization to mean any item in a collection. That is, it might be a work of art, a specimen, a document, an artefact etc.
6. American Institute for Conservation of Historic and Artistic Work, *Guidelines for Practice*, 18, revised 1994.
7. Richard Handler, "On the valuing of museum objects," *Museum Anthropology*, Volume 16, Number 1 (1992): 21–28.
8. Handler, p. 24.
9. Handler, p. 27.
10. It goes without saying that all members of any community may not share the same opinion. Likewise 'society as a whole' is not homogeneous, encompassing, for example, educational, income, gender, age as well as cultural differences. Often situations exist, though, where the representatives of a minority group disagree with whom they see as representing power held by the majority.
11. Sarah Harding, "Cultural Property and the Limitations of Preservation", *Law and Policy*, Volume 25, Number 1, (January 2003): 17–18.
12. Harding, p. 18.
13. David Leigh, "Closing Remarks, IIC Congress 2006, The Object in Context: Crossing Conservation Boundaries," as adapted in *IIC Bulletin* 5, October 2006: 2.
14. American Institute for Conservation of Historic and Artistic Works (AIC), "Section III" *Code of Ethics*, revised 1994.
15. Sarah Staniforth, "Conservation: Principles, practice and ethics," National Trust, *Manual of Housekeeping: The Care of Collections in Historic Houses Open to the Public* (Elsevier, 2006) 35.
16. Staniforth, p. 35.
17. Cited in: Christian F. Feest, "Repatriation: A European View on the Question of Restitution of Native American Artifacts," *European Review of Native American Studies*, Volume 9, Number 2 (1995): 33–42; and Ruth Phillips, personal communication, cited in: Miriam Clavir, "An Examination of the Conservation Code of Ethics in Relation to Collections from First Peoples," *First Peoples Art and Artifacts: Heritage and Conservation Issues. Art Conservation Training Programs Eighteenth Annual Conference, Professional Papers*, ed. K. Spirydowicz (Kingston, ON, Canada: Art Conservation Program, Queen's University, 1992) 1–2.
18. Cruikshank 25, quoting James Clifford, in an interview with Brian Wallis, "The Global Issue: A Symposium," *Art in America*, July 1989: 153.
19. Cruikshank, p. 28.
20. Cruikshank, p. 35.
21. Cruikshank, p. 31.
22. Nancy Marie Mithlo, "'Red Man's Burden': The Politics of Inclusion in Museum Settings," *American Indian Quarterly*, Volume 28.3 & 4 (Summer & Fall 2004): 745.
23. Mithlo, pp. 744–745.

24. Mithlo, p. 745.
25. Zvi Goffer, "The Causes of Decay," *Archaeological Chemistry: A Sourcebook on the Application of Chemistry to Archaeology* (New York: John Wiley and Sons, 1980), p. 239.
26. Sarah Staniforth, "Group Report: What Are Appropriate Strategies to Evaluate Change and to Sustain Cultural Heritage?," *Durability and Change: The Science, Responsibility, and Cost of Sustaining Cultural Heritage* (report of the Dahlem Workshop on Durability and Change: The Science, Responsibility, and Cost of Sustaining Cultural Heritage, 6–11 December 1992, Freie Universität, Berlin), eds., W.E. Krumbein, P. Brimblecombe, D.E. Cosgrove and S. Staniforth, 218–823. (New York: John Wiley and Sons, 1994) 218.
27. See also David Lowenthal, "The Value of Age and Decay," *Durability and Change*, eds., W.E. Krumbein, P. Brimblecombe, D.E. Cosgrove, and S. Staniforth (eds) (New York: John Wiley & Sons, 1994) 39–49.
28. Debra Sparrow, personal communication, 1999.
29. Staniforth, p. 35.
30. Susan Pearce, *Archaeological Curatorship* (Leicester: Leicester University Press, 1990) 106.
31. Miriam Clavir, *Preserving What is Valued: Museums, Conservation and First Nations*, (Vancouver, UBC Press, 2002) 43, referring to Susan Pearce, *Archaeological Curatorship* (Leicester: Leicester University Press, 1990) 106.
32. See for example: Mark S. Phillips, and Gordon Schochet, eds., *Questions of Tradition* (Toronto: University of Toronto Press, 2004).
33. Christopher Rowell, "The historic house context – the National Trust experience", National Trust, *Manual of Housekeeping*: *The Care of Collections in Historic Houses Open to the Public* (Oxford: Butterworth-Heinemann Elsevier, 2005) 8–19.
34. Jessica S. Johnson, Susan Held, Kelly McHugh, Elizabeth Brown and Marian Kaminitz, "Practical Aspects of Consultation with Communities," *AIC Journal*, Volume 44, Number 3 (Fall/Winter 2005): 203–215.
35. Ruth B. Phillips, "Re-placing Objects: Historical Practices for the Second Museum Age," *The Canadian Historical Review* 86, 1 (March 2005): 83–110.
36. Renata Peters and Dean Sully, "Finding the Fallen: Conservation and the First World War," *The Object in Context: Crossing Conservation Boundaries*: *Contributions to the Munich Congress, International Institute for Conservation of Historic and Artistic Works*, eds., David Saunders, Joyce Townsend and Sally Woodcock (London: IIC, 2006) 12.
37. Michael M. Ames, "Counterfeit Museology," *Museum Management and Curatorship* 21 (2006): 171–186.
38. Ames, p. 172.
39. Mithlo, p. 746.
40. Ames, p. 172.

14

The Cultural Dynamics of Conservation Principles in Reported Practice

Dinah Eastop

Introduction

This chapter provides a material culture analysis of conservation principles in practice by analysing how principles are invoked when reporting conservation practice. The underlying hypothesis is that conservation is a social process, and as a component of culture, it is open to different interpretations. A material culture approach integrates understanding of the material and social aspects of things, and facilitates an analysis of conservation as both a social and a technical practice, mediated by language. Material culture studies are concerned with why things matter to people, by seeking to understand the relationships between persons and things in the past and in the present.[1] The focus on the social role of things leads to analysis of the materials, technology and circumstances of an object's making (production), its use (or consumption) and its disposal. In this chapter, conservation interventions are analysed as dynamic social processes involving the interrelationships of people, things and language (Figure 14.1).

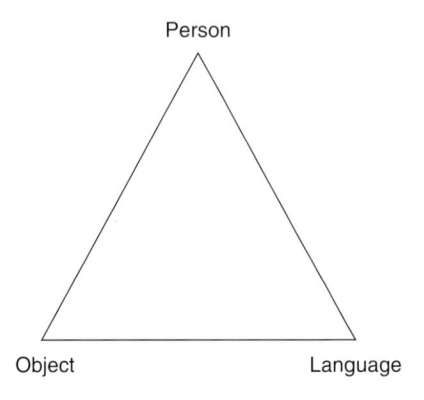

Figure 14.1 *Diagram representing the author's view of 'material culture' as the interrelationship between persons, objects and language.*

For the following analysis it is important to consider the range of meanings given to the term 'principle' and the way these meanings arise through the work of metaphor. The origin of the word principle is the Latin word *principium*, meaning first in time order, often emphasised as 'first principles.'[2] The three different but inter-related meanings of principle are origin, fundamental assumption, and rudiment (Figure 14.2). In the first, it means the original state, that from which something originates, a basic or fundamental source, or a primary element, force or law that produces or determines particular results. In the second, it means a fundamental truth, proposition or assumption forming a chain of reasoning. In this sense it forms a law or rule as a guide to action. Its third meaning is rudiment or the first part of study. Each of these meanings depends on a building-based metaphor, where the foundation (which has to be laid down first) provides the physical support for the rest of the structure. The work of metaphor allows for a transfer of meaning from the domain of building to the domain of principles. As fundamental assumptions, principles may be 'taken for granted' as the basis (foundation) of argument or action. Another important factor to note is that, in terms such as 'high-principled,' the word principle often carries morally positive connotations.

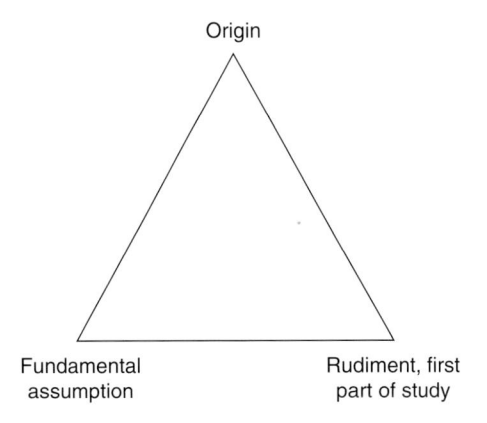

Figure 14.2 *Diagram representing three different definitions of principle.*

Conservation is defined here as preservation, investigation and presentation,[3] and may be viewed as part of production or consumption depending on the analyst's views of the aims, results and context of the intervention. This analysis is based on *The Object in Context: Crossing Conservation Boundaries*, contributions to the 21st Congress of The International Institute for Conservation and Restoration of Historic and Artistic Works (IIC), held in Munich in 2006.[4] This publication was chosen because it is international, and reflects a wide range of conservation specialisms; it has multiple contributions, which were selected and edited by senior members of the profession; it is a recent publication and so may be supposed to be topical; and it is widely available (in both hard copy and as a CD), so the papers are readily available for critique of the following analysis. It was presumed that the congress theme, how context affects conservation decisions, and 'crossing conservation boundaries,' would require recognition of professional boundaries and negotiation in the application of conservation principles.

Case studies

Three papers have been selected as the basis for this analysis. They reflect a variety of conservation specialisms and institutional roles while encompassing a range of material and object types: painted wood wall panelling; a fish trap made from plant materials; bronze and teak musical instruments. The following analysis focuses on the relations between people (in social organisations), objects (undergoing conservation interventions) and the language used to describe and explain the interventions. The importance of language in conservation has been highlighted by Laura Drysdale, who demonstrated how 'taken for granted' assumptions are expressed in the way conservation is reported.[5] Her paper also provides a model for analysing contributions to an IIC Congress. In the following analysis, each of the selected accounts is summarised; the tacit and explicit principles are then identified and analysed by reference to the different meanings of the term 'principle.'

In *The Berlin Aleppo Room: a view into a Syrian interior from the Ottoman Empire*, J.M. Schwed[6] describes the importance, display and conservation treatments of painted wood wall panels and cornices removed from a house in Aleppo. *The Berlin Aleppo Room* is the name given to a rare set of panelling, made between 1600–1603 for a merchant's T-shaped reception room in Aleppo, one of the most important and cosmopolitan cities of the Ottoman Empire, and now in northern Syria. The decorative scheme combines scenes from Christian iconography with court scenes and quotations from the Koran. The panelling was purchased in 1912 for the Museum für Islamischer Kunst (Museum of Islamic Art) in Berlin, where it is one of the museum's highlights.

The paper describes changes in the way the panelling has been displayed since 1912. Initially, a few panels were exhibited as exemplars of Islamic art. In 1932, most of the panels were displayed lining the walls of a rectangular exhibition room; the aim remained 'to present the panels as objects of art rather than to convey their relevance to an architectural context.'[7] The panelling was damaged and dispersed during the Second World War. It underwent a major interventive conservation treatment in 1960, which allowed the panelling to be displayed in a T-shaped configuration (Figure 14.3). The treatment of 1960 involved cleaning, removal of varnish, some restoration of the wood, fixing polychromy on the cornices, 'neutral retouching' and re-varnishing.[8]

Figure 14.3 The Berlin Aleppo Room *(Inv. No. I 2862), as displayed after the 1960 conservation. Photo credit: Georg Niedermeiser, Museum für Islamischer Kunst, Berlin.*

In 2006, the panelling was undergoing further conservation treatment for a new display. Recent condition assessment revealed that the wood is generally sound, but the paint is vulnerable due to the variety of grounds and binding media, conditions of low relative humidity and the embrittling effect of the 1960 consolidation treatment. Current conservation measures include materials analysis and documentation; improving the environmental conditions of the display area, which requires temporary restrictions on visitor access into *The Berlin Aleppo Room*; and planning the dismantling, packing, transport and reassembly of the panelling for its new display. Parts that have never been exhibited before will be shown, and lintels will be moved to their original positions; a reconstruction of the room's central fountain will be added to the display. The overriding aim is to help 'provide a better understanding of the room in its entirety'[9] and 'to experience the unique, genuine atmosphere of a comfortable oriental reception room.'[10]

There is no explicit mention of principles in Schwed's account of *The Berlin Aleppo Room*, but four tacit principles become evident. The current intervention focuses on investigation and documentation, on preventive conservation, such as environmental control, and on minimal intervention. Interventive measures have been limited to paint consolidation: 'intensive retouching is not intended.'[11] The fourth is the privileging of authenticity; it is presumed that an authentic experience is inherently desirable. The underlying principle is the primacy attributed to the room's original state, as a merchant's reception room, with concern shown for the effect that previous displays may have had on the viewers' perceptions of the room. The account, which starts with the historical importance and rarity of the panelling, stresses the conflicting demands of preventive conservation and of public access, in the sense of the museum's visitors getting a sense of the original atmosphere of the room.

The first definition of 'principle,' that of original state, is not only the rationale for the conservation intervention, but it is also applied in the conservation process. The conservation intervention acts out the material basis of the metaphor underpinning the term principle when the presumed original form of the panelling is privileged. This presumption is used to critique previous presentations and to explain the current interventions. The metaphor of foundation is used to justify a conservation approach, which is then used to re-establish the panelling's presumed original form, which is justified by the foundational metaphor of principle. Within this linguistic framework the current conservation of *The Berlin Aleppo Room* can be understood as enacting a series of interlocking metaphors, a process known as the 'play of tropes.'[12]

The conservation treatment and the rationale given for the intervention presume and enact the foundational metaphor of the first definition of principle, that of original state. In the current re-presentation, priority is given to presenting the panels and cornices in a form that will resemble as closely as possible their presumed original appearance. The account takes it for granted that returning the room to how it may have looked originally is the right thing to do. In this sense the argument for presenting the room in its original state manifests the second meaning of 'principle', where a fundamental proposition or assumption (or what is presented here as a self-evident truth of the primacy of the room's original form) informs the chain of reasoning for the current interventions. Thus the meaning of principle implied in the description of the conservation of *The Berlin Aleppo Room* is built by merging two different meanings of principle: original state and fundamental assumption.

In *Gamelan: can a conservation-conceived protocol protect it spiritually and physically in a museum?* H. Jones-Amin, H. Tan and A. Tee introduce a set of written guidelines for playing the gamelan in a museum[13] (Figure 14.4). They describe the materials and production of gamelan orchestras, which are fundamental to Javanese performing arts. Gamelan are attributed a divine origin and sacred power; they are highly revered and given individual names. In 2000, the Asian Civilisation Museum (ACM) in Singapore acquired a gamelan, which was made in *circa* 1960 and bears the name *Ngambar Arum*, for the ACM's Southeast Asian Performing Arts Gallery. The museum selected twenty-one pieces of the orchestra (made of teak and bronze, decorated with painting and gilding) for permanent display and for performance.

Figure 14.4 *The gamelan (*Ngambar Arum*) as displayed and played in the Southeast Asian Performing Arts Gallery, Asian Civilisation Museum (ACM), Singapore. Photo reproduced with permission from Heidi Tan, Asian Civilisation Museum, Singapore.*

Jones-Amin *et al.* discuss conservation issues arising from playing gamelan, such as the need for periodic re-tuning. Tuning is an irreversible process involving filing, hammering and the addition of mud, wax or other materials. 'The ethical implications of this interventive approach to tuning are measured against the traditional methods of gamelan maintenance and its contextual use as a musical instrument.'[14] The capacity to play the gamelan was viewed as more important than retaining it in its 'as acquired' state. Mechanisms were sought to acknowledge the physical properties, spiritual attributes and music-making capacity of the museum's gamelan, while also respecting its current context in a museum. For example, the pieces on permanent exhibition are displayed on a low platform to evoke the sacred space of gamelan performance, and an audio-visual display shows the gamelan being played. When interviewing gamelan makers and musicians, it was established that the spirit of the gamelan would be unaffected by the museum context, provided that the gamelan was accorded respect. A written protocol (which is outlined in the paper) for the care and use of the gamelan was developed as part of the museum's conservation strategy; this includes guidelines on respectful behaviour, and the use of incense and food offerings.

There is no explicit reference to principles in the paper, but the protocol is one means of achieving a compromise between preserving 'conceptual integrity'[15] and material integrity while acknowledging the physical changes resulting from the practices of tuning and playing the gamelan, and demonstrating respect for it. The foundational metaphor of 'underpinning' is used to explain both the material and the spiritual integrity of gamelan: 'The time spent learning gamelan and interviewing makers and players had a profound affect on the conservators' understanding of the underpinning materials and spiritual beliefs.'[16] The account starts by considering the effect on revered objects of changes in context and the development of the protocol to guide all users of the gamelan.

In *Conservation of a Māori eel trap: practical and ethical issues*, C. Smith and H. Winkelbauer describe the treatment of a large net trap, discovered in 1869 by a land surveyor, and acquired by Otago Museum, Dunedin, New Zealand in *circa* 1919.[17] The trap has a knotted mesh made from a local plant material, and was on display between 1982 and 2002. It became very fragile, brittle and distorted; loss of the plant material resulted in voids in the mesh (Figure 14.5). The initial conservation recommendation was to place the eel trap in long-term storage, as a form of passive conservation, because the conservator believed that an interventive treatment would put the fragile mesh at risk. In addition, the ethos of minimal intervention made in-filling the voids seem undesirable. Finally, the museum did not have the necessary staff time and expertise for such work.

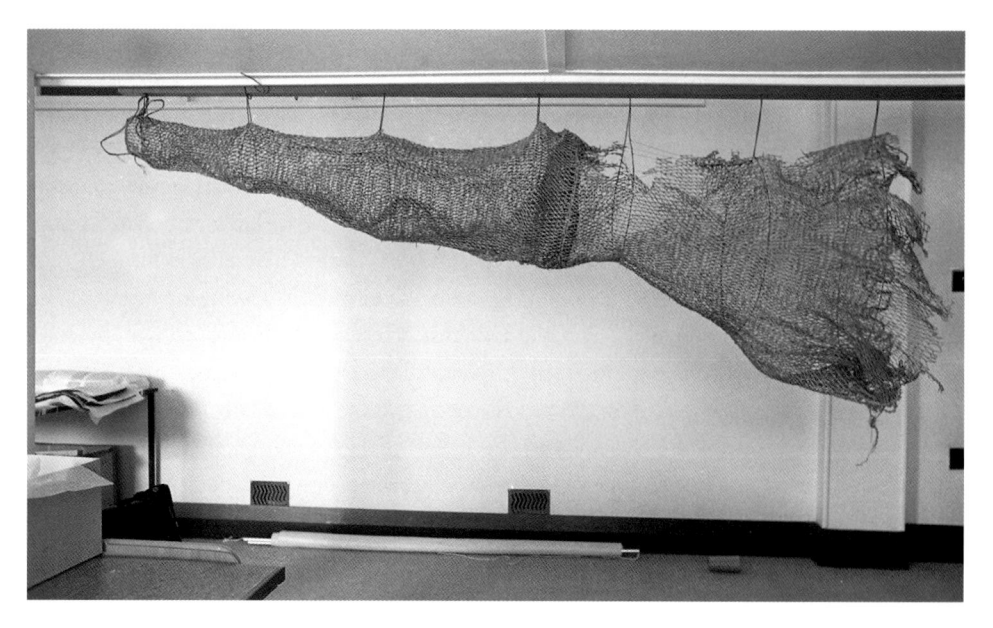

Figure 14.5 *A large eel trap found in Central Otago, New Zealand in 1869, before recent conservation interventions. Photo credit: Otago Museum, Dunedin, New Zealand.*

However, boxed storage was considered inappropriate for an object that is viewed as *taonga*, objects, places or activities to which Māori attribute qualities of power, fear and ancestral authority, and this led to a new conservation proposal. The museum's Māori Advisory Committee (MAC) stated a preference for 'repair and display in a manner that recognised its cultural value [to representatives of the originator culture].'[18] This meant that the 'damaged appearance of the eel net was clearly not fitting.'[19] This led to the implementation of an interventive treatment. The net was surface-cleaned and humidified in order to make it flexible enough to re-align the misshapen mesh; tears and breaks in the mesh were closed and secured with Japanese paper and 'conservation grade' adhesive. In-fills, made by a Māori weaver in the same knotting technique but in 'conservation grade' materials, were adhered to fill remaining voids in the mesh. As the remedial conservation work progressed, a funnel-shaped mount was constructed to support the net. Problems with gauging the size of the net meant that the mount was too big and one of the voids could not be closed before the exhibition. The eel trap was put on display despite the remaining void because it had 'a sound, cared-for and valued appearance.'[20]

The explicitly stated principles of passive [preventive] conservation and minimal intervention provided the basis for the initial conservation proposal. This proposal was not considered appropriate by representatives of the originator culture, whose principles of respect required the eel trap to be displayed and to have a sound and cared-for appearance, necessitating the in-filling of voids. The custom-made display mount fulfilled the preventive conservation needs of support, while the custom-made in-fills manifested the change from minimal intervention to restoration while respecting conservation principles. 'The use of in-fill material constructed from conservation-grade materials was deemed appropriate for two reasons: the long-term stability of the material, and easy identification of in-fill material as a non-indigenous repair.'[21] The choice of materials also accords with the principles of reversibility and re-treatability, and with the '6 ft:6 in rule' by making the in-fills obvious on close inspection but not from a distance.

The account of the eel trap's treatment draws on the second meaning of 'principle:' a proposition or assumption forming the chain of reasoning. The conservator's initial assumption was that care of the eel trap was best demonstrated by storage and minimal intervention. The assumption informing the interventive treatment was the view of the Māori *iwi* (tribal group) that appropriate care was demonstrated by displaying the trap with a valued appearance. Both interventions are based on a common principle of respecting an object's integrity; the differences arise in the way that notions of integrity and respect are applied to objects (the eel trap in this case) and are understood by people representing the cultures of origin and of conservation. The account starts by identifying different views of object integrity and the need for community consultation. The code of ethics of the New Zealand Professional Conservators Group (NZPCG) 'formally recognizes the primary role of Māori in regards to *taonga*;'[22] the Māori Advisory Committee of the Otago Museum provides one mechanism for such community consultation.

Material culture analysis

The anthropologist Daniel Miller argues that 'humans order things and are ordered by things.'[23] This idea is elaborated by Ingold: 'people not only bring order to things . . . they are also ordered by things, perceiving the world in accordance with the framework of meaning embodied in their artefacts.'[24] The dialectic relations of ordering and re-ordering are shown in the case studies.

The account of *The Berlin Aleppo Room* is based on a sequence of re-orderings. These include dismantling the panelling in Aleppo and exhibiting it in Berlin, and a display history which encompasses the 1930s display in a rectangular room, the 1960 reconstruction of the T-shaped room and the current proposal to re-position the lintels. The conservation interventions have also involved re-ordering, for example in removing varnish. Thus curatorial and conservation decisions have helped to create *The Berlin Aleppo Room* and to present the panelling in various forms. The ordering of people by objects is also demonstrated. In a literal sense, visitors have been placed in different relations to the panelling over the years. The current environmental controls mean that visitor access is restricted, and the proposed re-display will control the visitors' viewing point.

The Berlin Aleppo Room also has an ordering effect in the conceptual sense. Taken-for-granted assumptions (which often work as un-stated principles of the second type) about wholeness (closeness to presumed original form) reinforce ideas about authenticity, and its association with wholeness. This principle is articulated in the stated goal of 'helping to create a work of art that is understandable in its entirety.'[25] Certain conservation principles are reinforced by practice, and practice reinforces certain principles: in this case in attributing the object's 'true nature' to its original form. The influential role of curators and conservators in presenting certain views of objects is shown by the contrasting images on the cover of *Conservation in Context*. It shows the *Landsdowne Leda* with and without additions, demonstrating changing views of authenticity and completeness, and the changing aesthetic of the fragmentary and the whole.[26]

The dialectics of physical and conceptual (re-)ordering are also evident in the account of the gamelan. Physical re-ordering included moving the gamelan to the museum in Singapore, dividing the set of instruments so that part is kept in store and part is on display, the renewal or substitution of materials, such as crystalline wax for mud, and the filing and hammering of re-tuning. Re-ordering of people is manifested in the protocol, with respect for the gamelan shown by not stepping over the instruments. The effect of the gamelan on conceptual ordering is also shown by the respect accorded to the gamelan as an agent and by the fact that it was made an exception at the museum in being used for performances. This necessitated changes in preventive conservation measures, which usually focus on protecting physical integrity but were modified so that the gamelan can function as revered musical instruments. One modification was allowing offerings of incense and food to be brought into the museum. This led to rethinking of the principles of minimal intervention and of preventive conservation to encompass the conceptual integrity of the gamelan as an active, respected, named agent, and as part of an active performing arts culture.

The gamelan case study also provides an example of the compromises required to meet the sometimes conflicting principles of preventive conservation and promoting public access to collections. One effect of the protocol, which was developed to reduce risk of damage to the musical instruments during performance, is to modify the social context of play itself. Players cannot bring their snacks with them, and the restricted space in the gallery draws attention to the players rather than to the gamelan.

The conservation of the eel trap also involved re-ordering the object, for instance in the re-shaping of the mesh and the filling of the voids. The technical and ethical challenges posed by its conservation demonstrated the modified relations of custodianship arising from an institutional commitment to community consultation. It also led to changes in assumptions about the benefits of passive intervention in storage compared to interventive conservation for display. In this way the conservation of the eel trap led to changes in thinking about the appropriateness of minimal intervention and in 'taken-for-granted' concepts of wholeness. The paper demonstrates the expanding notion of an object's integrity, from physical state and history to include spiritual and cultural values attributed by representatives of the originator community.[27] The conservation of the eel trap is an instance where two value systems, or sets of principles, may be said to collide.[28] The resulting ethical and technical challenges help to explain why recognising the rights of originator communities is an important feature of recently written or revised professional codes of conservation.

Language (as in such professional codes) is one of the ways people order the world and are ordered by it. The bodily experience of self and of contact with things affects language development through metaphor. For example, the meaning of 'hard to understand' is derived from what we experience as physically hard (say, bumping into something hard) through the meaning transfer of metaphor from the sensory to the conceptual domain.[29] The metaphors that underlie the meanings of 'principle' have an effect on how conservation practice is understood and implemented. 'Principle' as the first or the base has the effect of privileging a [presumed] original state over other physical states, including alterations or additions, or social processes (such as mechanical functioning or ritual use). The building-based metaphor of the word and of the concept 'principle' has the inherent effect of privileging the material over the social (i.e. the object over the person in Figure 14.1).

This helps to explain why the principles governing the current proposal for conserving and presenting *The Berlin Aleppo Room* can be taken-for-granted. In contrast, the social demands and physical effects of musical performance necessitate articulation of principles because the treatment of the gamelan is seen as going against the normative principle of material integrity, based on a metaphor firmly rooted in the idea of material solidity (the foundations of a building). Explicit articulation of principles is required in the case of the eel trap's conservation where respect for the object's integrity led to two different conservation proposals – at opposite ends of the intervention spectrum. The physical integrity of the eel trap preferred by representatives of the originator community was a consequence of respect, rather than the reason for respect. The metaphorical underpinning of the English word 'respect' depends on looking and looking again. The metaphorical base of the Māori term translated as respect is likely to have different connotations. The assumption that the value of objects rests with their fixed, material forms reinforces the materials-based foundation of both the word and the concept of principle, producing a re-enforcing cycle.

Another taken-for-granted assumption in the account of *The Berlin Aleppo Room* is that the reader will agree with the principle of revealing or re-presenting the room in its (presumed) original form. There is no need to present the case, because it is self-evident as the basis for the argument and the account of the actions (both current and planned). To be successful, all assumptions in an argument remain silent or 'taken-for-granted,' i.e. they don't invite questioning. As principles may be understood as taken-for-granted assumptions, there is an obvious tautology here. All successful assumptions are circular, tautological and sustain a self-justifying ideology. Most of the papers, for which the primacy of the original state is a self-evident truth, draw on an art-historical paradigm in their accounts of conservation interventions. Those papers that demonstrate an understanding of objects where alterations, additions and maintenance (such as tuning the gamelan) are accepted as part of an object's true nature (integrity) are more likely to be debated in the rationale for interventions; examples include Kruger Grossman;[30] Thompson and Elliott;[31] Thorn.[32]

Conclusion

The case studies provide vivid examples of conservation as material culture where the dialectic relations between the ordering of things, people and concepts is played out. They show how principles are invoked and deployed in two main ways: to govern action or guide practice, and as ideology when reporting actions viewed retrospectively. Principles are used and/or invoked differently in different cultures and political contexts; some are stated explicitly, while others remain implicit. This affirms the idea that principles are culturally constructed, and that principles constrain and help to manifest culture (as part of social process):

- Principles can be 'taken-for-granted' when they encompass dominant or uncontested ideology.
- Principles are debated when power is more contested or egalitarian.
- Naming principles allows the principles and practices to be questioned and tested by the evidence.

- Principles are invoked as a higher order than practice because they are considered more abstract. However, as shown above, relations between principles and practice are dialectic, resulting in a process of circular feedback.

A material culture approach fosters recognition that the principles or ideology of conservation may be debated or taken-for-granted in a particular context with its own political realities. More specifically, it shows that social context is likely to influence which principles are invoked and how they are deployed. It also questions the apparent neutrality of conservation principles. The core argument is that the principles and the practice of conservation are cultural phenomena that are constrained historically and socially.

Acknowledgements

For help with securing images, granting permissions, and processing images, I thank: Otago Museum's Māori Advisory Council (MAC), Catherine Smith, Heike Winkelbauer and Scott Reeves; Holly Jones-Amin, University of Melbourne and the Asian Civilisation Museum, Singapore; Professor Haase, Director, and Jutta Maria Schwed, Museum für Islamischer Kunst, Berlin; Mike Halliwell, Textile Conservation Centre (TCC). I also thank David Goldberg for rigorous critique of earlier drafts, and Nell Hoare, Director of the TCC, for permission to publish.

Notes

1. C. Tilley, W. Keane, S. Küchler, M. Rowlands and P. Spyer (eds), *Handbook of Material Culture* (London: Sage, 2006).
2. *The Shorter English Oxford Dictionary on Historical Principles* (London: George Newnes, 1933).
3. D. Eastop, "Conservation as Material Culture", eds., C. Tilley, W. Keane, S. Küchler, M. Rowlands and P. Spyer, *Handbook of Material Culture* (London: Sage, 2006) 516.
4. D. Saunders, J.H. Townsend and S. Woodcock (eds), *The Object in Context: Crossing Conservation Boundaries* (London: IIC, 2006).
5. L. Drysdale, "The language on conservation: Applying critical linguistic analysis to three conservation papers", ed., J. Bridgman, *Preprints of the 12th triennial Meeting of ICOM's Conservation Committee, Lyon* (London: James and James, 1999) 161–165.
6. J.M. Schwed, "The Berlin Aleppo Room: A view into a Syrian interior from the Ottoman Empire", eds., D. Saunders, J.H. Townsend and S. Woodcock, *The Object in Context: Crossing Conservation Boundaries* (London: IIC, 2006), 95–101.
7. Schwed, p. 96.
8. Schwed, p. 97.
9. Schwed, p. 97.
10. Schwed, p. 100.

11. Schwed, p. 100.
12. J.W. Fernanadez, *Persuasions and Performances. The Play of Tropes in Culture* (Bloomington: Indiana University Press, 1986).
13. H. Jones-Amin, H. Tan and A. Tee, "Gamelan: Can a conservation-conceived protocol protect it spiritually and physically in a museum?" D. Saunders, J.H. Townsend and S. Woodcock (eds), *The Object in Context: Crossing Conservation Boundaries* (London: IIC, 2006), pp. 138–143.
14. Jones-Amin, Tan and Tee, p. 141.
15. M. Clavir, "Preserving conceptual integrity: ethics and theory in preventive conservation", eds., A. Roy and P. Smith, *Preventive Conservation Practice, Theory and Research* (London: IIC, 1994) 53.
16. Jones-Amin, Tan and Tee, p. 141.
17. C. Smith and H. Winkelbauer, "Conservation of a Māori eel trap: Practical and ethical issues", eds., D. Saunders, J.H. Townsend and S. Woodcock, *The Object in Context: Crossing Conservation Boundaries* (London: IIC, 2006) 128–132.
18. Smith and Winkelbauer, p. 129.
19. Smith and Winkelbauer, p. 130.
20. Smith and Winkelbauer, p. 131.
21. Smith and Winkelbauer, p. 131.
22. Smith and Winkelbauer, p. 130.
23. D. Miller, "Artefacts and the meaning of things", ed., T. Ingold, *Companion Encyclopedia of Anthropology. Humanity, Culture and Social Life* (London: Routledge, 1994) 399.
24. T. Ingold, "Introduction to culture", ed., T. Ingold, *Companion Encyclopedia of Anthropology. Humanity, Culture and Social Life* (London: Routledge, 1994) 335–336.
25. Schwed, p. 97.
26. E. Risser and J. Daehner, "A Pouring Satyr from Castel Gandolfo: History and conservation", eds., D. Saunders, J.H. Townsend and S. Woodcock, *The Object in Context: Crossing Conservation Boundaries* (London: IIC, 2006) 190–196.
27. Clavir, p. 53.
28. A. Herle, "Museums and First Peoples in Canada", *Journal of Museum Ethnography* 6 (1994): 62.
29. G. Lakoff, *Women, Fire and Dangerous Things. What categories reveal about the Mind* (Chicago and London: The University of Chicago Press, 1987).
30. A. Kruger Grossman, "Keeping it together: conservation, context and cultural materials", eds., D. Saunders, J.H. Townsend and S. Woodcock, *The Object in Context: Crossing Conservation Boundaries* (London: IIC, 2006) 1–6.
31. M. Thompson and A. Elliott, "The Mimbres journey: how shifting contexts necessitate a multi-disciplinary conservation approach", eds., D. Saunders, J.H. Townsend and S. Woodcock, *The Object in Context: Crossing Conservation Boundaries* (London: IIC, 2006) 116–122.
32. A.Thorn, "*Tjurkulpa*: A conservator learns respect for the Land, the People and the Culture", eds., D. Saunders, J.H. Townsend and S. Woodcock, *The Object in Context: Crossing Conservation Boundaries* (London: IIC, 2006) 133–137.

15

Why Do We Conserve? Developing Understanding of Conservation as a Cultural Construct

Simon Cane

Introduction

Why do we conserve objects? I have often found myself having to answer this question as a conservator, as a teacher and as an advocate for the profession. I never seemed able to construct responses that, from my perspective, were robust enough or wholly convincing, because nagging away at the back of my mind I felt that I really didn't have the intellectual framework or structure to support my ideas. I have found, over the years, that I am not alone in this thought.

I find that thinking about the principles and ethics of conservation is a challenging, enjoyable and often frustrating process. Teaching and talking about these issues is never short of fascinating and rarely fails to throw up new ideas and positions. Writing on these issues, however, is a different matter. One becomes immediately aware of the labyrinthine connotations and possibilities that can lead from what, at first reading, sounds like a simple statement, such as 'all treatments must be reversible' or 'preserving the true nature of the object,' or that 'all objects are of equal value and should be treated as such.' It was ideals such as these that formed the central tenets of conservation philosophy for conservators of my own and previous generations. We upheld them, almost unquestioningly, because we believed that they would protect and empower us as professionals. Despite developments in conservation theory that successfully challenge these ideals, they seem to persist as if they are in some way sacrosanct. This has resulted in frustration and a level of confusion throughout my career with the level of debate and discussion around the principles of conservation. The physical understanding of objects, what they are made of, how they are constructed and why they deteriorate have underpinned the frameworks and belief systems that the conservation profession has developed. These have, however, become manifest in an ideological structure that remains focused on highly subjective and virtuous ideals such as truth, authenticity and stasis. Whilst these axioms are being challenged, they remain as central tenets of much of the conservation discourse.

This fundamentally ideological position causes a number of problems, as the majority of conservation writing and research is based upon understanding phenomena through the use of science that can often contradict this ideological stance. The corollary of this is that the modern conservation movement has built its systems of value and understanding upon a confused and conflicting philosophy, as research reveals that change is a constant, and that any intervention in the form of treatment alters the physical state of an object. This in turn challenges the ideals of truth and authenticity because truth is dependent upon perspective and is therefore contentious. A truly authentic state is unattainable due to the process of physical change over time, the phenomena of decay, and the decision to intervene to slow the process of change. It is important to note that a number of conservators (such as Dinah Eastop, Miriam Clavir and Salvador Muñoz Viñas, amongst others) have also begun to challenge traditional thinking and develop discourse about the conceptual value changes that conservators can make through intervention. The development of that discourse is, I believe, constrained by the potential for 'group think' by conservators whose education, training and belief systems are firmly and exclusively rooted in the understanding and prevention of physical change. I believe that this focus has resulted in a self-limiting discourse developed around the efficacy of conservation at the micro/physical level. Laura Drysdale has highlighted the potential limitations of this approach in her paper, 'The Language of Conservation,' in which she discusses the limitations of the conservation lexicon and the impotence of a discourse based upon efficacy.[1] This discourse is, therefore, by definition not equipped to begin to deal with issues of value on a metaphysical or, indeed, social level, which results in a weakened foundation for the development of arguments about the need for, and value of, the conservation of cultural and historic material. There is, then, a need for the development of a conservation philosophy that acknowledges and considers the position and function of conservation within the systems of objectification and value that have developed out of the museum model. The museum model is significant here because it is out of the museum that the conservation profession and its practices have largely developed.[2] The analysis of the development of conservation and object theory are essential steps in developing a model that illustrates the function, position and value of conservation. I will, however, limit the discussion here to considering ideas of value, developing understanding of cultural models and attempting to develop a model that illustrates the position and defines the function of the conservation process.

Cultural value

A debate around defining and understanding the public value of cultural heritage has developed over the last few years. Challenging essays on the subject have been produced by John Holden, a member of an independent think tank called

Demos, which aims to influence the development of policy and practice in a number of areas, including culture, through generating debate and discussion.[3]

The UK government has used hard quantitative indicators to measure the impact of museums but these do not make any judgement on the quality of the visit or interaction. The work of Demos and others has persuaded Demos policymakers of the need for 'soft' qualitative indicators that measure the type, quality and impact of an interaction which is evidenced through the introduction of Generic Social Outcomes (GSOs). It is, however, the quantitative indicators that form the core of performance assessment in the cultural sector.

We have also seen the development of value indicators in the field of conservation through the development of ideas and models, such as cost-benefit (quantitative),[4] condition surveys (qualitative)[5] and, latterly, risk (qualitative and quantitative).[6] These are all attempts to produce numbers in one form or another to support the need for conservation or prove the need for improved management of the cultural asset in question. What they do not do is help us really understand what benefit society accrues from the conservation and preservation of an individual item or collection. That is not a question that we can answer easily. But we all, I hope, are fairly sure of the benefits of, say, the conservation of a research collection in terms of the contribution it can make to understanding big issues such as the environment. On an individual level there is the positive, sometimes life-changing impact that a conserved piece of art may have, or the conservation of a specific group of objects, such as those from the S21 Prison in Phnom Penh in Cambodia, where people were tortured and summarily executed under the regime of Pol Pot. These now act as evidence of that brutality and as an educational resource used to teach children about this dark and disturbing episode in their country's history (Figure 15.1).

The systems that society has created to objectify, control and interpret culture are cultural constructs: constructions made by society that define the structure and systems by which culture is organized.[7] We can use this idea of the construct to develop a model that begins to define the function of conservation. The aim is to create a broader philosophical foundation from which it may be possible to construct a more complex understanding of the function and value of the conservation process. This, in turn, should enable the development of a more challenging discourse that allows us to move outside of the restrictions of the objective physicality that have traditionally restrained the dialogue of conservation.

Conservation as a cultural construct

'Cultural construct' is a term that describes how a society organizes ideas, beliefs and values into a system or 'construct' that are understood and then accepted and adopted by that society. Historic and art objects exist in a cultural construct

Figure 15.1 *Torture Room at S21 Prison. Phnom Penh, Cambodia.*

in which they attain both cultural and financial value, as demonstrated through the world trade of art and antiquities. It is through this construct that culture becomes commoditized, and society ascribes agreed values to cultural material. Those values are dependent upon a range of agreed factors, such as rarity, demand, source, artist or relevance to a specific community or place. How value is ascribed depends upon from where one is making a judgement and what the 'norms' are for one's construct. For example, the judgement of the value of aboriginal material by a traditional Western museum may well be different and at odds with that of the aboriginal people from which the material originated. The use of the term 'value' is, therefore, loaded and highly subjective.

Museums are also cultural constructs that have been developed to perform a complex range of functions that reflect upon, and are used by, the various societies in which they exist. They contain collections of objects that are normally beyond the commoditized, mercantile constructs of the market. These collections are still ascribed value as it is the museum that holds objects, which are considered of value by society. Once objects have entered the museum, their value, however, is not usually based upon the realizable, commoditized value that the object would reach if it were to be sold. Instead, the value rests upon where the object fits within the scheme of existing collections, its research value and, in some cases, the number of visitors that it will generate. This is especially true for a museum displaying ancient Egyptian material that is not always of high value, or indeed particularly rare, but is enduringly popular with visitors.

Inside a museum collection, the decision to 'conserve' it is taken for a variety of reasons. The received view of museum conservators and indeed the general public is that objects are treated because they are perceived as being in need of treatment to make them durable, but this is rarely the only reason for an object coming to the attention of the conservator. The fact that an object from a museum collection receives conservation treatment is rarely to do with increasing monetary value, as it often is in the private, commoditized, sector. This is because museum collections are effectively 'beyond value' in the commoditized sense. It can be argued that the conservation process is commensurate with an increase in cultural value as the decision has been made that this object is now valued enough to undergo the process of conservation.

The conservation of historic cultural objects is a fairly modern idea that has developed within the cultural construct that is the museum system. Others have discussed the development of conservation[8] and it is generally accepted that it is a twentieth-century phenomenon that developed and expanded in Western museums in the second half of the century. This places the construct of conservation within the postmodern era with its complex and changing interpretations about ideas of culture and society. This, I suggest, has some bearing on how it has developed, as it could be argued that the conservation process is as much about producing objects that match perceived ideas about the world as they are about producing objects that reflect truth or authenticity.

As suggested above, the treatment of objects that are still commoditized, existing in the private sector, has differed fundamentally from the treatment of objects in the museum construct. The conservation treatment of an object outside of the museum is often based upon the calculation:

Z (Value of conserved object) > X (Value of object before conservation) + Y (Cost of conservation)

The realizable value of the object after conservation treatment should be more than its original value plus the cost of conservation treatment. This may be changing as people seek to conserve their own objects and collections for reasons other than increasing financial value. This trend is indicated by the success of such initiatives as the UK's 'Conservation Register,'[9] which provides a list of conservation practitioners to the public via the Internet, and an increased public interest in history and conservation. Table 15.1, developed by the author, is an attempt to define the various constructs that are necessary for conservation of historic cultural material to exist. The elements listed on each side suggest a construction in whole or part that by its nature either enables or obstructs the idea of conservation as practiced in museums in the West during the later half of the twentieth century. This construct reflects a conservation practice that is about perceived ideas of cultural value and is based in scientific theory and practice rather than those values that define the object as commodity, curiosity, religious relic or icon.

Table 15.1 Taxonomy of Cultural Constructs

No Conservation	Conservation
Wonderment	Taxonomy
Real	Hyper-real
Everyday	Art–Culture/Museum System
Not Collected	Collected
Private (Individual)	Public (Collective)
Commodity (Durable)	Beyond Value
Archaic/Classicism	Postmodern

Once the factors on the right side of Table 15.1 are in place – which reflects the cultural construct of the postmodern museum – then conservation can exist and function. Without these constructs there would be no requirement for conservation to exist; objects may be repaired and maintained but not conserved. It is worth noting that cultures of collecting do not exist in all societies. In those societies where it does not exist, there is no need for, or interest in, conservation of material culture for its own sake or for the benefit of others in the society, though it is arguable that intangible culture is conserved through other systems such as ceremony, song, dance and storytelling. In this model the construct of the museum and conservation are interdependent, their values and aims entwined and mutually beneficent. This view, it could be argued, does not take into account the development of museums in countries that have little or no tradition of museums, and it should be noted that, whilst many take their lead from established Western museums models, they are developing in ways that reflect their own culture. They will, however, create their own constructs and conservation will, and is, appearing within those constructions often endorsed and supported by long-established European or American museums. The construct of the museum, then, provides the required environment for conservation to function and histories of the development of conservation confirm this to be the case.[10]

The cultural construct of the museum is where the conservation of a supposed intrinsic historical value of the object is the primary consideration over and above any financial value. The conservator also has a role to play in the investigation of the object to enhance knowledge and understanding, and these actions also preserve and enhance cultural value. Treatment decisions are usually based upon what the conservator considers best for the object given what it is made of, its condition and its intended use. The acts of the conservator are not, however, benign or neutral, and are influenced by the construct of the museum within which they operate, a point that has been noted by others, such as Eastop and Cosgrove:

> (Textile) conservators are active in prioritizing one history over another.[11]

> Any decision to deploy specific technical skills to restore an object to its 'original' state, to an intermediate state, or simply to keep it from further change is by definition arbitrary and should be recognized as such.[12]

The art-culture system

To help to gain an understanding of the position and function of conservation, it is necessary to consider the complex cultural systems in which museums and objects exist. It is helpful to this discussion to consider the idea of the 'Art-Culture System.' This is a model by which we attempt to define the structure of the system or construct that governs the movement of cultural material and the changes in value attached to that movement. The use of models to aid in the understanding of complex systems of value is quite common in social and cultural theory and Michael Thompson has developed a useful model, which acts as a starting point for this part of this discussion. His 'Rubbish Theory' is a 'dynamic system of cognitive categories,' that helps to illustrate the movement of objects and the accompanying shift in their value (Figure 15.2). In his 'Rubbish Theory,' Thompson proposes three categories of objects – transient, rubbish and durable – and describes the ways in which objects can move between these categories.[13]

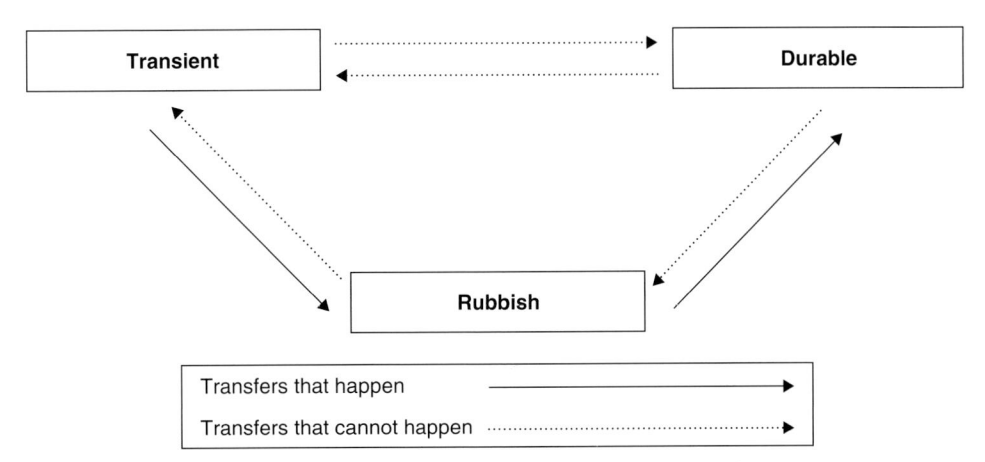

Figure 15.2 *'Dynamic system of cognitive categories' (Reproduced with permission from Thompson,* Rubbish Theory, *p. 45).*

A simple illustration of this is a piece of archaeology, such as a hand tool, which has been recovered from a historic waste site, a common source of material for archaeologists. When the tool was complete and functioning it was in the 'transient' category, a viable commodity. When it was broken it moved to the 'rubbish' category as it had been discarded and had no value. Finally, the tool is recovered, conserved and put on display in a museum, now it is in the 'durable' category where it is likely to remain. The museum, however, is placed at the limits of, or beyond, the durable category, but generally objects must pass through the durable category to enter the museum.

> One limit which durables undoubtedly approach, and which some classes of durable items actually attain, is total removal from circulation[. . .] The complete transfer of a class of items to museums and public collections is consonant with a general belief that, if only those items were in circulation, they would be increasing in value. In other words, they are so durable they are priceless.[14]

Whilst this model is useful in illustrating where the museum construct fits, and therefore where conservation can preside within the construct, it is rather simplistic and is unable to relate the complex relationships that exist within the Art–Culture System. James Clifford has developed a more complex model, 'a machine for making authenticity,' that helps us to understand how cultural material moves within the system, and that these transfers can, and do, result in a change in value (Figure 15.3).[15]

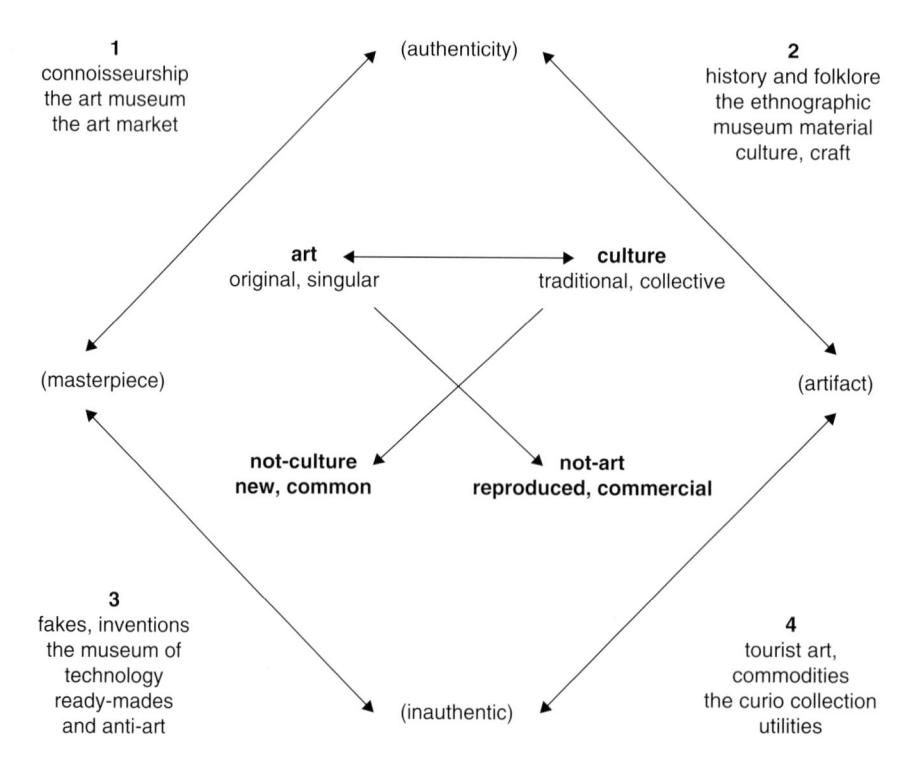

Figure 15.3 *'A Machine for Making Authenticity' (Reproduced with permission from Clifford, p. 224).*

In his model, Clifford places the four zones in opposition creating horizontal and vertical axes. Objects can be located in a specific zone or ambiguously in transit or oscillating between zones. Objects that move from the bottom (inauthentic) to the top (authentic) are regarded as demonstrating a positive movement, i.e. a rise in cultural, artistic and/or monetary value. Clifford illustrates the idea of value change of objects, describing how, in the 1940s, the Surrealists purchased Kuskokwim Eskimo masks from the Museum of the American Indian and placed them on exhibition, thus shifting their value from being objects of science to objects of art.[16] This example demonstrates how objects can be reclassified and re-valued as trends and thinking change. Conservators could be better equipped to respond to value changes such as these, but ideas around equality in value persist in conservation thinking and literature. This model provides a framework for the movement of objects within the Art-Culture System and introduces a number of rules about how and where objects can move. Clifford points out that any such system is by its nature procrustean and represents a historical view of a system in a state of constant flux.

A question that arises when considering these systems is if, and how, the conservation process affects the movement and value of objects. As discussed earlier, Thompson places the museum object beyond the durable category as he defines objects as being out of circulation, and therefore beyond value. This would suggest that the conservation process within the museum is outside of this model. Perhaps, then, it is more helpful to consider an Art-Culture System as proposed by Clifford, rather than just the museum, as being beyond Thompson's model. This is also problematic, as objects within the 'Art-Culture System' proposed by Clifford are still commodities, fitting in the upper end of Thompson's durable category until they enter the construct of the museum. Here they are effectively priceless, beyond value, their value being in what they are or what they were collected to represent and, importantly, owned by everyone and no one.

To help illustrate and understand how the conservation process may interact within systems, such as those proposed by Clifford and Thompson, I propose a model that attempts to map the movement of cultural material and illustrates where the conservation process fits within these constructs (Figure 15.4).

This model separates the art-culture construct and the museum construct, and they are distinct and separate from the everyday/non-art-culture zone. Objects can move directly from the everyday/non-art-culture zone into either the art-culture *or* museum constructs. Objects leave either of these constructs for conservation and a movement such as this would usually result in a positive change in value. This would usually be an increase in financial value in the art-culture construct as it overlaps with the durable, commoditized value system. For example, an antique is purchased by a dealer, undergoes conservation treatment and is sold for more than the cost of purchase plus conservation. Objects in the museum construct can make the same journey, leaving the museum and entering the conservation construct – often a notional journey if a museum has a conservation department, but one

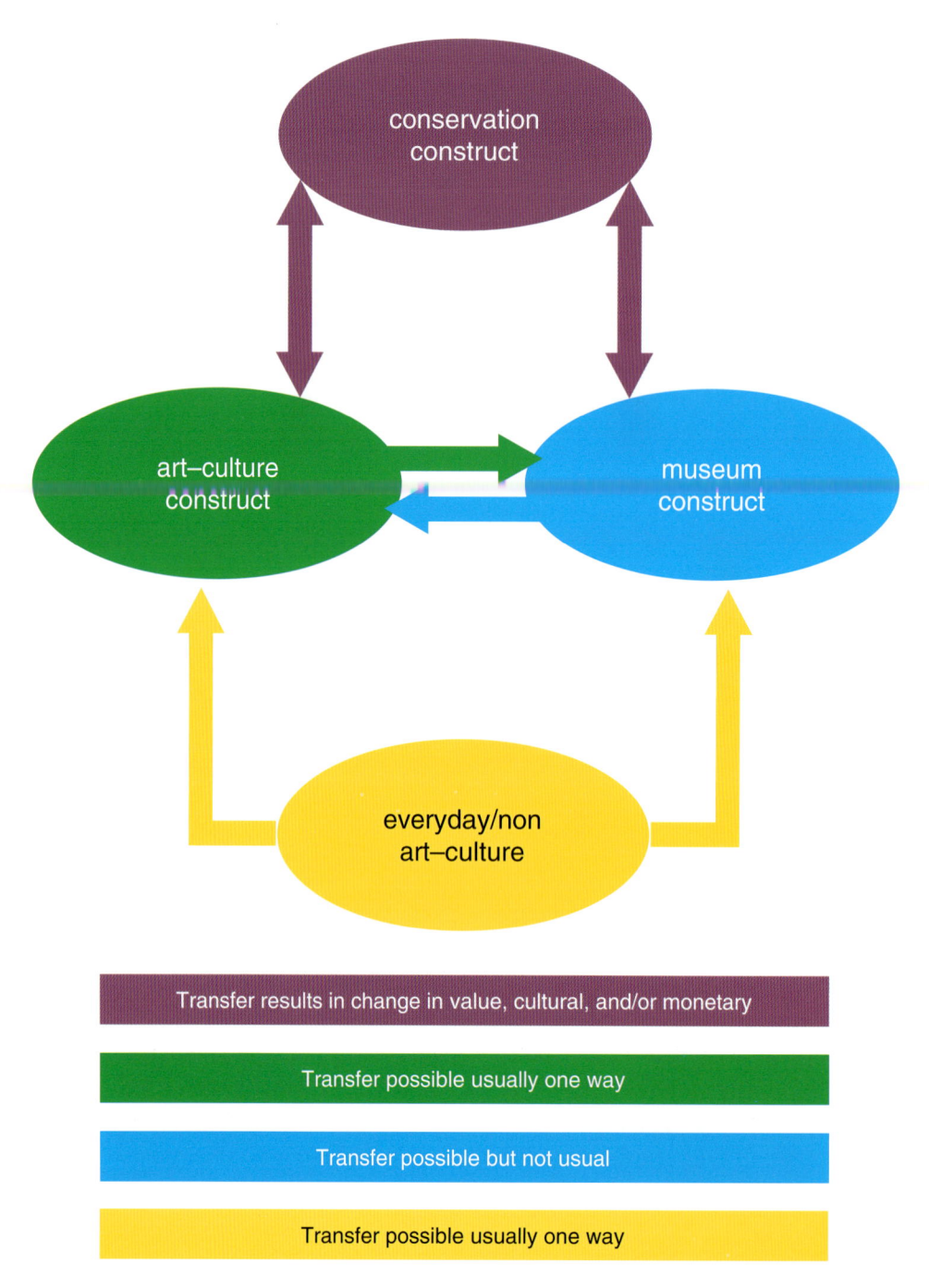

Figure 15.4 *A model of the inter-relationship of the conservation construct.*

that does involve movement of the object – importantly the value increase here is usually cultural and aesthetic. It is worth noting, however, that incorrect treatment could result in a loss of monetary and/or cultural and/or aesthetic value in either construct. It is, of course, true that the value of the object as commodity may well increase whilst the object is in the museum, but that increase in value is not realizable until the object leaves the museum and re-enters the commoditized world of the art-culture construct. In the model that I propose, this journey is possible but not usual. In North American museums, the idea of the object as commodity is more readily accepted and objects are regularly traded to enable new acquisitions. The trading, however, usually requires the object to re-enter the commoditized zone as sales are handled by a third, independent, party in the form of a dealer or auction house, people with experience in handling culture as commodity. Whilst museums in Europe will often acquire through auction, they very rarely realise the value of the commodities that they hold through the market. It is a matter of public trust and perceived integrity in the United Kingdom, as museums have a presumption against disposal of collections through sale or exchange, with any museum choosing to sell items from their collections facing vilification and criticism from public and professional alike. An object usually acquires increased cultural value just by entering the museum construct; the conservation process enhances and increases that value, as treatment of objects in this area is invariably linked to exhibition or the perceived importance of the object. Objects will often enter the art-culture construct and undergo conservation before moving to the museum construct. The conservation process in this case makes the object more appealing, appropriate and available. An example of this is when a painting is reattributed as a result of conservation work that makes the work desirable to a public art gallery or museum.

As with the models cited earlier, it has to be accepted that this model is somewhat rigid and offers an historical view of what is a complex system in constant flux. It does, however, help to illustrate that conservation is a necessary and important element in establishing and sustaining cultural value. Ideas and perceptions of value do change, resulting in objects being withdrawn from use and display, and not receiving conservation until their value shifts again and they are once again required to perform. The conservation of privately owned objects that sit somewhere between the art-culture construct and the everyday/non-art-culture zone is increasing, as evidenced through the plethora of printed and television media dedicated to the subject. The principle, however, that the conservation process represents a positive value change holds true, and the model could be modified to represent this.

Conclusion

The model developed by the author offers a means by which to locate the conservation process as part of a broader cultural construct, and opens up the possibility of developing more detailed and complex models to illustrate and understand the function and therefore the value of the conservation process. It is not proposed as being in anyway definitive. It allows for the engagement of conservation with the metaphysical ideas of cultural value rather than limiting it to a closed ideology of absolutes and so-called definitive truths that are, as pointed out earlier, problematic in developing a discursive discourse. Through the introduction of the metaphysical, we have another element of the equation, which may help deal with the dichotomy between the ideological aspirations of conservators, and the realities of our phenomena-based observations and understanding. The restricting consequences of a self-limiting, closed dialogue have been identified by John Holden as an issue for the cultural sector as a whole in his essay, 'Cultural Value and the Crisis of Legitimacy.'[17] Michalski has also highlighted the challenge for the conservators in his 1994 'Sharing Responsibility for Conservation Decisions' in which he describes the conservation discourse as 'the last bastion of that archaic narrative.'[18]

There are people, such as Salvador Muñoz Viñas, Miriam Clavir, Dinah Eastop, Jonathan Ashley-Smith and Robert Waller, who are challenging the ideological tenets that have dominated the conservation of the historic cultural heritage, and thereby enabling the development of a discourse that will hopefully lead us towards a dynamic contemporary theory of conservation. Training and education is also changing, as syllabi adapt to meet the demands of the changing work environment, and courses such as the ICCROM sharing decision-making programme bring disciplines together and encourage mutual trust and understanding. My own observation, however, is that there is a default setting to which conservators seem to want to return in which the 'conservation professional' is defined by a narrow, technical definition of skills and knowledge, as illustrated by recent attempts to develop a professional profile for conservators in Europe. This, I believe, reveals an underlying reluctance to engage in the broader, challenging debates around the vulnerability, care and use of the finite resource that forms the cultural heritage.

Conservators are facing an increasingly complex cultural environment where new ideas about ownership, decision-making and ethics challenge the traditional, largely Western, perception of the conservator as arbiter of aesthetic value, saviour of objects and keeper of scientific knowledge. The use of modelling, as suggested above, which has formed a small part of contemporary conservation thinking for some time, and the development of the understanding of conservation as a construct as well as a process, gives us the opportunity to respond from an intellectual and philosophical basis to those complexities. This will help in building a foundation from which to illustrate the need for, as well as the value of, conservation. My own experiences of museums and conservation practice over the last 25 years enforces my belief that the conservation profession needs to develop a robust and confident discourse that creates the intellectual space to adapt to new agendas that are developing around engagement, use and value. I asked the question 'why do we conserve?' at the beginning of this chapter, and I hope that I have gone some way to creating a map that can give some direction to the answer. Conservation is as essential as any other part of the processes of the museum in defining and maintaining cultural value, but this is not necessarily universally acknowledged. Conservation can serve to reveal hidden information, it can increase the lifetime of an object, and enable increased enjoyment and understanding of cultural material. Conservation of cultural material is an intrinsic part of our need as a society to collect, organize and display culture. It is a complex intellectual and physical process that raises many ethical, technical and philosophical challenges; and it is through those challenges that we gain a better understanding of the world as it was, as it is and, to some degree, will be.

Collections and other cultural material are under increasing pressure due to globalization, the impact of environmental change, increased demand, diminishing resources and inconsistent policy provision. The conservation of cultural material is essential to the cultural health of society, but we cannot assume that those who legislate and manage our society take this as read. Conservators need to find new, innovative and relevant ways to illustrate value, and for that we must engage with others, including those who own and use the culture we conserve. By continuing to open up to, and engage with, other areas of thought, conservators can develop a more robust intellectual framework, which in turn will encourage the development of practice and ensure that our shared, valuable and fragile cultural heritage has a viable future. So in answering the question 'why do we conserve?' we must consider a wide range of factors, which include understanding how things are made, why they fall to pieces and our innate desire to attain a sense of authenticity and truth about our world. Our reasons for conserving cultural heritage are complex and constantly evolving but it is clear that if conservation did not exist we would have to invent it.

Notes

1. L. Drysdale, "The Language of Conservation: Applying critical linguistic analysis to three conservation papers", *12th Triennial Meeting Lyon 29 August–4 September 1999 ICOM Committee for Conservation Preprints*, ed., J. Bridgeland (London: James & James, 1999) 161–165.
2. A. Oddy and S. Smith, eds., *Past Practice Future Prospects* (London: British Museum, 2001).
3. J. Holden, *Capturing Cultural Value: How culture has become a tool of government policy* (London: Demos, 2004).
4. M. Cassar, *Cost/Benefits Appraisals for Collections Care* (London: Museums and Galleries Commission, 1998).
5. S. Keene, *Managing Conservation in Museums* (London: Butterworth-Heinemann, 1996) 136–158.
6. Robert Waller, *Cultural Property Risk Analysis Model* (Gottenburg: Acta Universitatis Gothoburggensis, 2003).
7. T. Ingold, "Introduction to Culture", *Companion Encyclopedia of Anthropology*, ed., T. Ingold (London: Routledge, 1994).
8. C. Caple, *Conservation Skills: Judgement, Method and Decision Making* (London: Routledge, 2000).

9. The Conservation Register: (www.conservationregister.com).
10. E. Pye, *Caring for the Past: Issues in Conservation for Archaeology and Museums* (London: James and James, 2001).
11. D. Eastop, "Textiles as Multiple and Competing Histories", *Textiles Revealed: Object lessons in historic textile and costume research*, ed., M.M. Brooks (London: Archetype, 2000) 26.
12. D.E. Cosgrove, "Should We Take It All So Seriously?", *Durability and Change: The Science, Responsibility and Cost of Sustaining Cultural Heritage*, eds.,W.E. Krumbein, P. Brimblecombe, D.E. Cosgrove and S. Staniforth (London: Wiley and Sons, 1994) 263.
13. M. Thompson, *Rubbish Theory* (Oxford: Oxford University Press, 1979) 272.
14. M. Thompson, "The Filth in The Way", *Interpreting Objects and Collections*, eds., S.M. Pearce (London: Routledge, 1996) 273.
15. J. Clifford, *The Predicament of Culture*; Twentieth-Century Ethnography, Literature and Art (London: Harvard University Press, 1996) 224.
16. J. Clifford, *The Predicament of Culture: Twentieth-Century Ethnography, literature and Art* (London: Harvard University Press, 1996) 239.
17. J. Holden, *Cultural Value and the Crisis of Legitimacy: Why Culture Needs a Democratic Mandate* (London: Demos, 2006).
18. S. Michalski, "Sharing Responsibility for Conservation Decisions", *Durabiltiy and Change: The Science, Responsibility and Cost of Sustaining Cultural Heritage*, eds., W.E. Krumbein, P. Brimblecombe, D.E. Cosgrove and S. Staniforth (London: Wiley and Sons, 1994) 252.

16

Heritage, Values, and Sustainability

Erica Avrami

Introduction

Over the past several decades, a significant dialectic has emerged within the conservation field[1] between the global and the local, the universal and the particular. This tension has challenged some of the fundamental ethics of conservation practice and compelled a re-examination of heritage and its value to society. Traditionally understood as a reified concept and body of resources, heritage and its conservation have the potential to evolve in the postmodern world as key contributors to sustainability. However, such development of the field requires new emphasis on the social processes of conservation and a reorientation of the underlying principles of practice.

The global-universal

In the years since the 1962 Venice Charter, the conservation field has produced dozens of transnational conventions, declarations, as well as documents pertaining to the protection and management of immovable cultural heritage.[2] Hand in hand with these has been the development of a global infrastructure of organizations, legislation, and programs. These collectively bear witness to the maturation of conservation as a legitimate profession and field of study.

A notable milestone was the founding of the International Council of Monuments and Sites (ICOMOS) by UNESCO in 1965. ICOMOS launched an international network of practitioners and academicians and laid the foundation for a common language of heritage conservation. The 1972 World Heritage Convention proved to be an equally seminal tool for safeguarding sites worldwide, and created a newfound solidarity amongst the national bodies responsible for conservation. The notion that some resources were of 'universal value' to all of humanity likewise fostered dialogue and cooperation across borders and sectors of society. The stewardship of the historic built environment was cast as a shared responsibility at the global level.

These combined international efforts have had a synergistic affect on conservation. Cross-cultural collaboration has enhanced education and research. It has set precedents and guidelines for national and local heritage policy and management. Such co-operation has also had an influential role in the 'standardization' of professional practice through shared principles about how to conserve.

The local-particular

Following on the heels of the globalization of conservation has been a growing recognition of the importance of local knowledge and public participation in heritage protection. This response is largely due to developments in planning theory and social movements in the second half of the twentieth century. As society transitioned from the modern to the postmodern era, planning and management of the built environment underwent significant paradigm shifts. Rooted in a long history of utopian thought that emphasized scientific knowledge, rational planning practices unraveled as the community consequences of many urban renewal efforts became evermore apparent. Concurrently, planning theory was increasingly challenged as part of the growing postmodern critique of Enlightenment epistemology.

Key aspects of this evolving discourse are derived from the work of Jürgen Habermas and other social theorists, who challenged the rationalist tradition and elucidated the function of deliberation and the free exchange of ideas in social action.[3] Their application to planning has promoted broader participation of stakeholders, challenged 'top-down' expert-driven models, and helped to transform planning into a more socially and contextually-responsive endeavor. There is now greater consensus in the field regarding inclusive dialogue in planning processes, the recognition of social difference, and increased understanding of the different ways in which knowledge is created and transmitted.

There are many significant forces in today's society that necessitate this reconceptualization of planning theory and practice, including international and rural-to-urban migration, postcolonialism, the resurgence of indigenous peoples, and the rise of organized civil society.[4] These forces have likewise had profound affects on conservation, which is one of the essential tools for managing the built environment. Repatriation, the rights of indigenous groups over particular sites, grassroots community-based preservation initiatives – all prompt new perspectives on how we conserve and why. Conservation is not simply about the objective stewardship of heritage resources, but is largely bound up in the very subjective relationships between people and places.

Stemming in part from the aforementioned theoretical developments and forces, 'value-driven planning' has thus emerged within the conservation field. Australia's Burra Charter was one of the early instances of such approaches to conservation being codified as part of a national policy. Hinged on the participation of a range of stakeholder groups and individuals, value-driven planning seeks broad public and professional input regarding decision-making about a heritage place or resource.

At the core of this planning methodology is a fundamental acknowledgement that values are ascribed to heritage by society at large. Values about what to preserve and how to preserve are derived from the meanings and uses that people attach to buildings, sites, and landscapes, and constructed amongst individual, institutional, and community actors. The values of certain stakeholders may conflict with those of others, and values may change over time or as a result of political conditions.

This more particular and temporal view of heritage and its significance gives greater weight to local knowledge and stakeholder perspectives. It highlights the very essence of heritage: that these resources differentiate one place from another, one community from another. Their uniqueness, because of associated meaning or added value, symbolizes the past of a particular society and helps to define the distinctive character of a locality. Conservation is therefore a fundamentally local act, though shaped by constituencies that may or may not be in close geographic proximity.

Tensions of the past

This global–local/universal–particular dialectic poses interesting challenges for conservation. On the one hand, the globalization of the field has served to legitimize the profession and practice of conservation and given rise to a community of experts and institutions who govern what to conserve and how. It has underscored the universal nature of heritage and fostered international co-operation on a range of fronts, from education to research to policy. It has likewise served to establish a common language of professional practice.

On the other hand, postmodern thought has engendered new questions and considerations *vis à vis* heritage and its cultural relativity. With the rise of value-driven conservation and the recognition of different ways of engaging with one's heritage, some of those universal ethics that have served to standardize practice are called into question. As Salvador Muñoz Viñas effectively argued in his 'Contemporary Theory of Conservation,' reversibility, authenticity, scientific objectivity and other long held tenets of the field are under challenge. Likewise, the role of conservation professionals is evolving. No longer are we simply experts prescribing an appropriate course of action, but also facilitators of a socially-responsive process that Muñoz Viñas refers to as 'negotiative conservation.'[5]

Conservation has traditionally been viewed as a neutral act of stewardship. It was premised upon a curatorial paradigm, underpinned by the principles of connoisseurship and involving expert identification of architecturally, historically, and culturally significant structures. However, conservation is not an impartial process of discerning some sort of intrinsic value. Rather, it is a creative process of valorizing a given resource or element within the built environment for the purpose of perpetuating a particular idea or narrative about a place or people. Decisions about how to conserve a resource or element likewise reflect the very complex ways in which places are significant to different people at different times.

More and more, scholarly analysis of the conservation process suggests its very subjective and political nature. As Glassberg argues, the collective memories ascribed to places 'emerge out of dialogue and social interaction,' but are likewise the consequence of 'conflicts with political implications over the meanings attached to places.'[6] In its most robust form, conservation is a tool for managing change and for codifying collective memory and storytelling in the built environment. The process of conservation can provide a vital means of community-building by reinforcing shared histories, cultivating collective identities, and fostering a sense of place. It can likewise serve as a dangerous vehicle for exclusion and ideologies of difference, and as a means of preventing, rather than managing, change.

This tension is exacerbated as populations become more heterogeneous, as knowledge production becomes more prolific, as the 'experience economy' thrives, and as globalization incurs rapid changes in social structures and landscapes.[7] The conservation field seeks to underscore the universality of heritage so as to promote cohesion within and across societies through a shared past. However, collective memories are rarely singular; rather they are precarious amalgamations of multiple narratives over time. Thus, the field is also struggling to recognize the particular voices – of the many individuals and communities – that contribute to such narratives by ascribing values to vestiges within the built environment.

Communicative and advocacy planning theories have informed and spurred the application of a more value-driven and deliberative process through which stakeholders can engage in the determination of what is heritage and how it should be safeguarded. However, stakeholders must still negotiate the institutional arrangements through which the politics of conservation play out. And ethnic and cultural groups that do not adhere to the prevailing theories of how to preserve or what is appropriate (which are based largely on Western European experience) are at a clear disadvantage in the participatory process. So while stakeholder values and participation have become part of the general rhetoric of practice, how cultural differences and multiple knowledge systems translate to a new set of principles for conservation remains largely uncharted territory.

Challenges of the present

These evolving social conditions in contemporary civilization bring added complexity to heritage conservation, but are likewise compounded by global concerns regarding 'sustainability.' Climate change, resource consumption, and population shifts – and their underlying economics – have made apparent, if not dire, the need to revolutionize the way we live in the industrialized world. The associated challenges of sprawl, construction waste, and energy use in buildings compel explicit transformations in how we design, construct, and manage the built environment. As communities try to combat the market pressures of urban growth, grapple with shifting demographics, adapt to the influx of immigrant populations, and apply sustainability principles to land use decision-making, conservation is becoming increasingly significant and controversial in the struggle to maintain continuity and manage growth, yet meet the demands of necessary change.

While the concept of sustainability has been framed as a tripartite of environmental, economic, and social factors, current debate focuses primarily on the former two. Consequently, the built heritage field has invested significant effort in recent years to articulate the economic and environmental rationales for conservation. The message that conservation can be both profitable and 'green' is resonating, and there is significant momentum in the field to build a body of knowledge that unequivocally supports this assertion. In the meantime, there has been only limited scholarship advancing the traditional mainstay of conservation: its relevance to social sustainability. It is clearly important to enhance assessments of the environmental and economic benefits (and costs) of reusing existing structures versus building new, but we cannot forsake the social implications as conservation engages more readily in decision-making about sustainable development.

Research demonstrating the complex social effects of heritage conservation has nonetheless proven difficult. While there is a fundamental conviction that heritage conservation benefits society at large, quantifying and qualifying the specific dynamics and outcomes is no easy task. The economist David Throsby has notably advanced this inquiry by demonstrating how heritage contributes to sustainability by generating tangible and intangible benefits, such as maintaining diversity, promoting inter- and intra-generational equity, and underscoring the interdependent nature of the cultural infrastructure of a place.[8] These and other examinations focus on heritage as a form of cultural capital, or essentially products of cultural practice that have created value. However, is the value of heritage solely bound up in these *products*? Or are the social benefits likewise, or even more so, generated by the *process* of conserving?

The fundamental significance of built heritage has long been vested in the 'place' – the building, the streetscape, the archaeological site, etc. By reifying the concept of 'heritage' in physical structures and landscapes, the conservation field has promulgated the notion that the social benefits of its efforts are embodied in the conserved place – or *product* – and society's experience of it. The presence of vestiges of the past within the built environment is essentially assumed to make us better citizens. However, little emphasis is placed on the social effects of the conservation *process* itself. Indeed, deciding what to conserve and how to conserve it may engender benefits that have less to do with the place itself and more to do with the way in which heritage serves as a potential vehicle for creating social and political capital. In other words, maybe the most significant contribution of heritage to social sustainability is the role of the conservation *process* in building community, recognizing differences, and enhancing social cohesion.

Opportunities of the future

Given the topic of this compendium, it is important to note that conservation principles have traditionally focused on how to treat the object or place, and on the *product* of intervention. However, as discussed previously, this mind-set has forced the field to confront some thorny issues in the postmodern world. Globalization incurs the need for cooperation and shared values, and conservation has met the challenge by promoting the universality of heritage and establishing a common set of professional ethics. But the very localized and political nature of heritage betrays a 'one size fits all' approach. Indeed, the very essence of heritage is its celebration of difference: certain places and structures are significant because people have developed associations and attachments to them that distinguish them from others. Accordingly, how to conserve such places – and the multiple narratives and values ascribed to them by various stakeholders – entails a complexity that extends well beyond the traditional tenets (i.e. authenticity, reversibility, etc.) that have guided intervention.

Though they may take a variety of forms and be approached in myriad ways, it is still nonetheless difficult to deny the universality of heritage and its conservation. But it is possible that the shared values incurred by globalization play out more in the potential social benefits of conservation, rather than in its physical outcomes. Indeed, what if conservation principles focused instead on the how people engage in decision-making about their heritage and the *process* of participation?

Such an approach to conservation has essentially been advocated through value-driven methodologies, though without explicitly articulating the underlying canons or guiding principles for the evolution of practice. Conservation is fundamentally a form of planning – both public and political – that seeks to codify collective memory in the built environment, so as to communicate the values of a community to future generations. Those values are often contentious and conflicted, the narratives layered and discordant. The conservation process is a potential vehicle for giving voice to multiple publics, encouraging deliberation, championing local knowledge, empowering communities, and negotiating change. Politics and power may dominate in such localized negotiations about heritage; the conservation process serves to mediate these relationships in an effort to find a common vision for the future through a collective past. In doing so, there is what Uffe Jensen refers to as a 'deference to an ideal of human flourishing' that transcends the particular and the local.[9] It is precisely this ideal that epitomizes the universality of conservation: through difference and deliberation we seek shared understanding.

Conservation is not merely an act of stewardship that privileges the past over the present; it is a creative destruction of alternative futures. A successful and sustainable vision for the future hinges on motivating human agency through broad public participation and accessible discourse. Heritage conservation provides a means to such ends, not simply because of the resources it safeguards, but because of the civic engagement it engenders. These political dynamics of conservation can promote social sustainability when deliberation is inclusive and underpinned by the fundamental principles of freedom, equality, and equity. Thus it may be that the essential aim in our work is to democratize the structures and processes of conservation so as to ensure these principles.

Notes

1. 'Conservation' in this paper refers to conservation of immovable heritage/the built environment.
2. See inventory of Cultural Heritage Policy Documents compiled by the Getty Conservation Institute (available at http://www.getty.edu/conservation/research_resources/charters.html)
3. Jurgen Habermas, trans. Thomas McCarthy, *The Theory of Communicative Action*, 2 vols (Cambridge: Polity, 1984).
4. Leonie Sandercock, *Cosmopolis II: Mongrel Cities of the 21st Century* (London: Continuum, 2005) 3.
5. Salvador Muñoz Viñas, "Contemporary Theory of Conservation," *Reviews in Conservation*, no. 3 (2002): 30.
6. David Glassberg, *The Place of the Past in American Life* (Amherst: Univ. of Massachusetts Press, 2001) 116–117.
7. National Heritage Board. *Towards Future Heritage Management: The Swedish National Heritage Board's Environmental Scanning Report* (Stockholm: The Swedish National Heritage Board, 2006) 16.
8. David Throsby, "Cultural Capital and Sustainability Concepts in the Economics of Cultural Heritage", ed., *Assessing the Values of Cultural Heritage*, M. de la Torre (Los Angeles: J. Paul Getty Trust, 2002) 101–117.
9. Uffe Juul Jensen, "Cultural Heritage, Liberal Education, and Human Flourishing," *Values and Heritage Conservation*, eds., E. Avrami, R. Mason, and M. de la Torre (Los Angeles: J. Paul Getty Trust, 2000) 43.

17

Ethics and Practice: Australian and New Zealand Conservation Contexts

Catherine Smith and Marcelle Scott

Introduction

Despite geographic proximity, similar colonial histories, and closely aligned social demographics, Australia and New Zealand present divergent cultural and political circumstances. These parallels and distinctions are also reflected in the conservation ethics and practices of both countries. A comparative study of the socio-political contexts of Australia and New Zealand was undertaken to show how these distinctive environments have influenced current approaches to the conservation of cultural material in each country. This approach was further informed by insights into the contemporary practice of conservation gained through a survey of members of the New Zealand Conservators of Cultural Material (NZCCM) and the Australian Institute for the Conservation of Cultural Material (AICCM), which sought opinions on the strengths, weaknesses and influences of conservation codes of ethics and practice from members.

The AICCM and NZCCM Codes of Ethics closely conform with the spirit, and in many cases the letter, of similar national and international codifying documents. In particular, each acknowledges the numerous values inherent in objects and sites. Both use the term 'cultural property' despite it being problematic. However, both codes differ from other international conservation codes through the inclusion of relatively recent amendments recognizing Indigenous peoples' particular privileges and responsibilities concerning the conservation of their cultural material. The recognition of the rights and wishes of Indigenous people, which has been instrumental in shaping policy and practice in conservation and the broader heritage sector, is of particular relevance to any discussion of conservation ethics in the Pacific region.

The NZCCM was formed in 1983, and adopted a Code of Ethics at its Annual General Meeting of 1985. The 1985 document was revised significantly in 1995 to formally recognize the primary role of Māori in regards to *taonga*. *Taonga* are 'all dimensions of a tribal group's estate, material and non-material.'[1] The

AICCM formed in 1973, and in 1985 adopted a Code of Ethics and Guidance for Practice that closely mirrored international codes, and would in turn inform the wording and intent of other codes as they were introduced. Various revisions of the AICCM Code occurred over time with the most significant amendments passed in 2000, recognizing the primacy of Indigenous peoples' rights and concerns regarding their cultural material, and stating that conservation practice must adapt to cultural requirements; the role of significance in conservation decision-making; and the need to minimize the impact of conservation activities on the natural environment. While it is noteworthy that both the NZCCM and AICCM introduced these important amendments, more telling is the date the amendments were adopted. New Zealand's recognition of the special relationship of Māori to their own material culture significantly predates the Australian amendment that formally acknowledged a similar role for people of Aboriginal and Torres Strait Island descent. In order to understand this divergence between the development of cultural materials conservation and its ethical precepts, an overview of the distinctive cultural and historic environments of each nation is required.

New Zealand and Australia – divergent cultural and historic contexts

Aotearoa New Zealand was settled from East Polynesia probably in the thirteenth century AD. While earlier European contact had been made (1642, Abel Tasman), it was James Cook's (1769) reports of the rich natural resources of New Zealand that led to the establishment of industries there in the late eighteenth and early nineteenth centuries. The lawlessness of related settlements, missionary activity, and a developing relationship with Britain, resulted in attempts to codify relationships with Māori.[2] The resultant Treaty of Waitangi (1840) between Māori and the British Crown established Māori as British subjects, with the same rights and protection under British law as Europeans already residing in Aotearoa New Zealand.[3]

In contrast, British colonization of Australia did not include a treaty with Indigenous people. In Australia, British colonization practices and subsequent legislation were based on the concept of *terra nullius*, the false notion that land was unoccupied at the time of British settlement in 1778, ignoring a well-defined system of land 'ownership,' use, and social practices that had been developed over thousands years. The absence of a formal treaty, and the concomitant denial of the human and land rights of Australian Indigenous peoples, can be seen as a major factor influencing the different socio-political histories of Australia and New Zealand.

The Treaty of Waitangi is considered the founding document of the modern New Zealand state, and frames New Zealand as a bicultural nation.[4] Biculturalism can be defined as 'a context where two founding cultures are entitled to make decisions about their own lives for mutual co-existence,' and is considered a contentious term by many.[5] The Treaty of Waitangi positions Māori and Pakeha (New Zealanders of British descent) as the two founding cultures in a national partnership, with equal rights to existence and governance. There are, of course, many other cultural groups who now live in New Zealand who have little status according to this national framework. In consideration of the multicultural reality of contemporary New Zealand, some consider the Treaty as between *tangata whenua* (Māori, the people of the land) and *tangata tiriti* (people who belong to the land by Treaty right), or in other words any other cultural groups aside from Māori residing in New Zealand.[6] Despite this, many New Zealanders feel that the cultural diversity of contemporary New Zealand society is inadequately represented in a bicultural framework.

Article II of the Treaty of Waitangi provides Māori with authority over their lands and *taonga*, and this has important connotations for an understanding of Māori views and the expectations of museum and conservation professionals.[7] The three-dimensional *taonga* commonly found in museums are not simply 'things'; rather they embody important aspects of Māori culture.[8] Contemporary Māori have living relationships with *taonga*, which can be the physical manifestation of an ancestor, and are also seen as part of the *whakapapa* (genealogy) of a tribal group that links members to other physical and spiritual resources, irrespective of the era in which they were made.[9] Even *taonga* held in museums with no tribal provenance, divorced from cultural knowledge and narratives (*korero*) by European collection, are valued by Māori and are seen as their direct responsibility.[10]

The Treaty of Waitangi, and the responsibilities it implies, has increasingly become a part of the New Zealand cultural landscape. Political and constitutional recognition of Māori culture and rights developed throughout the 1970s and 1980s (sometimes referred to as the *rangatiratanga* (sovereignty) movement) as did the idea that non-Māori New Zealanders had the responsibility to develop appropriate new, post-colonial relationships with Māori.[11] This bicultural environment, and recognition of the Treaty of Waitangi, acknowledges the particular relationship and responsibilities that Māori have towards *taonga*, and that *taonga* have special intangible and spiritual values best understood by Māori.

In Australia, it was not until the 1967 national referendum that Indigenous Australians were granted rights of citizenship. Another twenty-five years would pass before the notion of *terra nullius* was formally overturned. After a decade of legal action, the 1992 landmark *Mabo* judgement of the High Court of Australia gave legal recognition to Indigenous Australians' land title claims. Responding to the subsequent 1996 High Court decision in the *Wik* case, which confirmed that pastoral leases[12] do not extinguish native title, then Liberal Party Prime Minister John Howard commented that the 'pendulum had swung too far towards Aborigines and had to be reset'[13] and, in 1998, the Native Title Amendment Act was passed. While not without vocal critics in public and political spheres, the *Mabo* and *Wik* judgements can be seen as accurately reflecting the views of a large sector of the community. The Reconciliation movement of the late 1990s and early 2000s saw public participation in marches and other manifestations of public opinion on a scale not seen since the Vietnam/American war, and not seen again until the 2003 protests against the invasion of Iraq. The popular underpinning of the Reconciliation movement was finally made apparent through Prime Minister Kevin Rudd's delivery of a formal apology to Indigenous Australians on 13 February 2000. This apology for past wrongs had been a central platform of the Labor Party's successful election campaign.

This brief overview of the colonial and recent political histories of Australia and New Zealand provides a background from which a better understanding of the social and political influences that have shaped museological and conservation policy and practices can be gained. As Ken Gelder and Jane Jacobs state, ' . . . Indigenous claims for sacred sites and sacred objects over the last twenty years . . . [are] crucial to the recasting of Australia's sense of itself.'[14] This 'recasting' of Australia's sense of self has led to changes in heritage policy, and for many conservators has been central to their practice and professional values. Regrettably, employment of Aboriginal and Torres Strait Islanders in conservation has not been 'recast,' and in matters related to Indigenous cultural heritage the profession in Australia must continue to consult outside of itself rather than be informed from within. In direct contrast, a Treaty partnership and a bicultural framework in New Zealand have contributed to the training and employment of Māori in museums, including as conservators, and has arguably positioned Māori in a far more influential role in the heritage sector. The existence of professionally trained Indigenous conservators in New Zealand has profoundly influenced the foundation and ideological underpinning of conservation principles, and the development of the conservation profession.

Implications of divergent socio-cultural contexts on heritage policy and conservation in New Zealand and Australia

The growing voice for self-determination over *taonga* ignited by the internationally toured exhibition *Te Māori*, as well as the Māori cultural renaissance, was also distinguished by concerns over the conservation and preservation status of valued *taonga* held in museum collections. The importance of *taonga* in a living Māori culture meant that Māori articulated concerns not just about control, but also the preservation status and conservation of *taonga* held in museum collections: 'to be guardians we need to commit to conservation.'[15] From a very early point in the history of professional conservation in New Zealand, the special nature of *taonga*, and the importance of preserving its spiritual and cultural significance in museums, were also recognized by non-Māori, along with the recognition that to do so appropriately and effectively required Māori knowledge and input.[16] This recognition was formally codified in Article 4 of the NZCCM Code of Ethics:

> Māori customary concepts empower particular knowledge of heritage and conservation values to chosen guardians, with respect to particular places and artefacts . . . all members of NZCCM shall recognize the special relationship of Māori to places and artefacts as described in the Treaty of Waitangi.[17]

While the spirit of biculturalism is evident in the literature on New Zealand museums, no legislation requires museums in New Zealand to be bicultural. Of the four major museums in New Zealand[18] only one (Auckland Museum) makes direct reference to the Treaty in its governance framework.[19] However, the Acts of these museums all provide for some measure of input from Māori, and the Museum of New Zealand Te Papa Tongarewa (the National Museum of New Zealand) has, from its planning and inception, actively pursued bicultural policy and practice, and employed Māori staff at all levels.[20] Gerard O'Regan's (1997) assessment of biculturalism in museums showed that many were not truly bicultural.[21] In the words of David Butts (2003), 'institutional biculturalism is applied like makeup'; in reality changing nothing of the underlying issues of colonialism and cultural appropriation in museums.[22] Overall, the level of biculturalism at each cultural institution seems to be determined by its historic relationship with local Māori, and how that relationship has been fostered. Many smaller cultural institutions in New Zealand have more meaningful and equal partnerships with Māori, reflecting a serious commitment to Treaty obligations.[23]

Formal recognition of Indigenous peoples' rights in the Australian museum sector came much later than that in New Zealand, and was closely aligned with, and influenced by, changes in public sentiment and reconciliation processes occurring since the 1980s. In 1993, the Council of Australian Museum Associations (CAMA) issued *Previous Possessions, New Obligations*, the first national policy statement to inform museum practices for the care and management of Aboriginal and Torres Strait Islander cultural heritage.[24] The guiding principle of the document was the 'recognition of the inherent interest of Aboriginal and Torres Strait Islander peoples in the care and control, spiritual and practical, of their cultural property' and in Principle 8 stated that 'Conservation practice must adapt to cultural requirements.'[25] Many museums and individuals in Australia had been changing their practices over the preceding decade or longer. Nonetheless, the formal and public acknowledgement of the primary rights of Indigenous people in the management of their cultural heritage was influential, as was the consultative process that led to the formulation of the policy. In 1995, Australia became one of the few countries worldwide to introduce a national conservation policy. The policy recognized that 'Museums have particular obligations to conserve and preserve the movable culture heritage of Aboriginal and Torres Strait Island communities and peoples.'[26]

In the opening plenary of the 1999 ICOM-CC Triennial Meeting, the then AICCM Vice-President flagged the intention of the Institute to 'recognize as a guiding principle the inherent rights and interests of Indigenous peoples in the care and control, spiritual and practical, of their cultural property.'[27] The influence of the CAMA policy is obvious in the wording of this statement. By 2000, when the AICCM welcomed the International Institute for Conservation (IIC) conference to Melbourne, the Australian profession had formalized its intent. Taking advantage of the presence of the international audience, the AICCM issued a public statement outlining the Institute's commitment to Reconciliation. The Statement recognized Aboriginal and Torres Strait Islander peoples' ownership, and right to self-determination in the preservation and representation of their material culture, and apologized for the injustices of the past, and their continuing consequences.[28]

Codes of ethics – Australian and New Zealand conservators' perspectives and practice

To inform a critical analysis of the respective codes of ethics, and to substantiate insights into the effect of amendments on current professional values and practices, the authors conducted a survey of members of the AICCM and the NZCCM. Of the seventy members who expressed interest in participating in the survey, twenty-nine returned completed questionnaires.[29] The survey sought to determine members' level of familiarity with the codes; views on whether or not the codes were representative; ways in which the codes were used; and their influence on the professional practice of individual members and the sector more broadly. In particular, the questionnaire sought to gauge members' views on the key amendments made over the last decade.

While the sample size was not sufficient to produce statistically relevant quantitative data, indicative trends, general views, and a 'snapshot' of the current professional context within which conservators in the region operate can be seen. The relatively low response rate from the membership, and the lack of published quantitative studies of the influence of conservation codes on conservators' views and practices, supports Frank Matero's (2000) assertion that the conservation profession has avoided a critical analysis of its historical and professional constructs.[30]

Survey results

Most respondents indicated moderate familiarity with the codes. Others added that they refer to the codes on a fairly regular basis, and were therefore very familiar with their content. There was near universal agreement that the codes accurately reflect the aims and objectives of the conservation profession, indicating the centrality of ethical frameworks to the practice of conservation.

Strengths and weaknesses of the codes

When asked about key strengths of the codes, respondents identified their importance as a framework for decision-making, and their contribution to professional credibility. Several commented that the recent amendments, which 'reflect changes in society and within the profession', were the strength of the codes.

Conversely, when asked to identify the key weaknesses of the codes, a small number of respondents felt they lacked sufficient detail to inform decision-making, and that the language was pedantic. The majority described the main problem as the lack of a clear process of enforcement.

The second key weakness identified related to the prescriptive, rigid nature of the codes, which some respondents felt 'held back change.' This was elaborated by one respondent who suggested that the 'lack of discussion of the ethics and context for responsible preservation of tangible heritage and use is trapping conservators in an old mindset, or at least failing to offer leadership for them to develop a new more sophisticated outlook.' A concern raised by several respondents, and linked throughout the survey, was that the highly prescriptive tone of the codes, along with their emphasis within education programmes, continually encourages new generations of professionals to adopt without question a narrow definition of their role. This has a direct implication for conservation education programmes and reinforces the onus on the discipline to engage more proactively with key documents and professional precepts.

Other comments related to concerns about the prioritization of the tangible over the intangible, which ignored values associated with use and function. Respondents also identified gaps in the codes related to recognition of maker's intent, the impact of the digital era on documentation and on the preservation of original format, and the need for a 'whole of collection' rather than a single-object focus. These responses appear to contest one of the universal principles of many international codes – an unswerving respect for the integrity of the object. Whether intended or not, this principle has historically been subject to a narrow interpretation, one that priviledged the physical object over its inherent meaning and is premised on the view that conservators are best qualified to determine an object's integrity. Well before this survey was conducted Alison Wain (2000) expressed similar concerns, stating:

> Conservators should not restrict themselves to looking after the physical aspects of objects – to do so places artificial barriers to the development of conservation as a broad and flexible discipline and alienates other cultural heritage professionals . . . Current conservation codes of ethics should be revised to explicitly define the preservation of content and function as legitimate and ethical aims, even where these conflict with preservation of physical material.[31]

These views suggest that explicit recognition of other values, beyond those based purely on scientific method and an emphasis on the physical artefact, is required to better reflect the changing contexts of culture and heritage. What may have once been seen as universal truths are now open to challenge, as assumptions about meaning, ownership and use of collections change.

Indigenous peoples' rights and wishes

The overwhelming majority of respondents indicated that amendments recognizing the particular rights of Indigenous peoples reflected an essential aspect of conservation practice, and that the inclusion of these clauses was of the utmost importance for raising awareness within and outside the profession. However, a few respondents (both New Zealand and Australian) indicated that consideration of Indigenous perspectives could be adequately addressed in other clauses of the codes relating to intangible values. In the Australian socio-cultural context this could be linked to the 'pendulum' comment discussed previously whereby Indigenous Australians were perceived as gaining unfair advantages over non-Indigenous Australians. In New Zealand there is increasing debate about biculturalism, which can be seen as privileging two sectors of the community (and implicitly Māori) over other groups in an increasingly multicultural society.

Despite the broad agreement in both countries regarding the importance of recognizing the rights of Indigenous peoples to self-determination in the conservation of their cultural heritage, there is wide variance in the level of actual engagement of Indigenous people in conservation policy, decision-making, and practice. In New Zealand, Māori are actively involved in the conservation of their heritage, an involvement that is explicitly recognized in the Treaty of Waitangi, and implicitly at central and local government level. Notwithstanding concerns raised over the actuality of the commitment to biculturalism in museums in general, Māori are clearly seen as both stakeholders and staff who have specific and necessary cultural knowledge. In contrast, the level of employment and active engagement of Indigenous Australians varies across the museum and heritage sectors, but in general remains very low. To simply blame broader factors of discrimination for this situation is complacent and inadequate. A systemic structural change of professional and workplace cultures and educational opportunities and pathways in Australia is necessary.

Archaeology and archaeological conservation in Australia does have a strong history of collaboration with Indigenous Australians, and this is clearly seen in the field of rock art conservation. Rock art sites of course remain in their original locations, requiring conservators to travel to these destinations and to work with Indigenous people on their terms. In museums, where consultation with Indigenous people and communities is often mediated through curatorial and other departments, true ongoing collaboration is less frequent, although some notable exceptions exist. In New Zealand, some conservators are more likely to have contact with Māori, and consult and collaborate as a recognized requirement of professional practice. These conservators are usually from those specializations concerned with Māori artefacts that have been historically valued, collected and held in museums (wood carvings, textiles, greenstone and bone artefacts). Working in institutions that are bicultural in approach, with strong *iwi*[32] relationships, facilitates active and effective consultation and collaboration, as does working with a conservation colleague who identifies himself or herself as Māori. As ideas about *taonga* broaden to incorporate a more diverse range of disciplines in contemporary art and collections practice, it is anticipated that all New Zealand conservation practitioners will need to engage in consultation processes, rather than just have familiarity with the concept.

This disparity in the extent of collaboration across specialization, geographic region, and organizational focus has resulted in considerable variation in the knowledge base and practices of conservators in Australia and New Zealand, and may also explain views expressed in the survey that the recognition of Indigenous peoples' wishes should be covered in other clauses that relate to 'physical, historic, aesthetic, and cultural' values. However the majority view of respondents was that the amendments provided a clear signal of the intent of practising conservators to recognize the particular rights of Indigenous peoples. Also interesting, however, were responses indicating that these amendments lagged behind professional practice, rather than providing guiding principles, as exemplified by the comment '[the amendment] has formalized/acknowledged the approach/beliefs of the majority of practicing conservators i.e. they were most likely doing it anyway.' This suggests that the desire for consultation and closer collaboration exists, even though the practice may be lacking in some areas.

Discussion and conclusion

Any discussion of Codes of Ethics must recognize that such documents are in essence a series of value statements and principles that seek to represent a consensus view designed to guide individual behaviour. This chapter has focused on the humanistic aspects of conservation, as they are represented in the Codes of Ethics and Codes of Practice of the profession in New Zealand and Australia. In doing so, we have sought to emphasize the link between the practice of conservation and the broader socio-political situations in those countries.

A survey of members of the professional bodies for conservators in New Zealand (NZCCM) and Australia (AICCM) provided insight into the contemporary practice of conservation in the geographic region. Respondents to the survey unanimously agreed there was an important ongoing role for the codes in describing the framework and standards of professional conduct. They stated the codes can facilitate 'technical and philosophical dialogue' and 'reflect contemporary ideals and aspirations of the community of conservators *as well as* the broader community' [authors' emphasis]. This latter outward-looking comment confirmed recognition of the profession's public accountability. The amendments recognizing Indigenous people's rights and wishes; environmental impact; and significance as a decision-making tool serve to demonstrate to ourselves, to our professional colleagues and to the public, the continuing growth and expanding field of influence of the discipline. The changes are critical and important in reflecting unique local and regional cultural differences in the Australian/Pacific region.

Respondents, however, felt that these amendments reflected pre-existing or emerging behaviours, and in that sense lagged behind, rather than influenced, professional opinion and practice. Therefore, while the NZCCM and AICCM can be seen as international leaders in regard to these code revisions, the survey showed that the amendments embody existing ideas and behaviours. New Zealand and Australian conservators have clearly stated their aspirations and expectations regarding the involvement of Indigenous people in conservation, there continues to be variance in practice across areas of conservation specialization; from one institution to another; in different regional areas; and from one side of the Tasman to the other. Additionally, codes in this region still retain consistency with international codes that exalt the physical over other values and place an emphasis on the single object over a broader focus.

This study found that there is a strong desire amongst professional practitioners in Australia and New Zealand for continual revision and further investigation of conservation codes of ethics to enable better understanding of the key issues that influence and drive the discipline. A narrow interpretation of codes of ethics, resulting from a lack of debate about their true nature, hampers critical evaluation of conservation theories and practices, limiting intellectual discourse, and restricting advances in approach. Critical examination of professional precepts is essential to progress beyond narrow, inflexible interpretations of values, and to adopt and promote truly inclusive practices. This requires a willingness to question existing paradigms and a preparedness to engage in broader dialogue outside of conservation about issues central to culture, heritage and humanistic concerns.

The amendments to the codes of ethics and codes of practice in New Zealand and Australia can be seen as a mandate to the professional conservation bodies of other nations to more publicly and directly align themselves with issues of import to Indigenous peoples' heritage. The view that the codes of ethics in Australia and New Zealand are more relevant as a result of these revisions suggests a strong acknowledgement by the profession of the numerous constituents it serves. Codes of conservation ethics need to be relevant, and their language and intent inclusive. As documents that exist to prescribe the values of the profession to members and to the public, it is highly desirable that regular critical review of the codes of ethics and practice takes place. This would encourage greater engagement with and critical examination of our professional tenets, creating guiding documents with the recognition and authority to play a dynamic leadership and aspirational role.

Acknowledgements

The authors wish to thank all members of AICCM and NZCCM who participated in the survey, and those who gave permission to be quoted in text. Aspects of Marcelle Scott's research were undertaken while a 2007 Conservation Visiting Scholar at the Getty Conservation Institute. She is grateful for research assistance and support from Valerie Greathouse and Judy Santos in the GCI Information Centre during this time. The authors would also like to acknowledge the assistance and support of colleagues from the Department of Clothing and Textile Sciences, University of Otago, Dunedin, New Zealand and the Centre for Cultural Materials Conservation, University of Melbourne, Australia.

Notes

1. Ian Hugh Kawharu, *Waitangi: Māori and Pakeha Perspectives of the Treaty of Waitangi* (Auckland: Oxford University Press, 1989) 321.
2. Michael King, *The Penguin History of New Zealand* (Auckland: Penguin Books Ltd, 2003).
3. Arapata Hakiwai, "The Search for Legitimacy: Museums in Aotearoa, New Zealand – A *Māori* Viewpoint", *Proceedings of the International Conference on Anthropology and the Museums*, ed., L. Tsong-Yuan (Taiwan: Taiwan Museum, 1995) 282–294.
4. Mason Durie, "*Māori* and the State: Professional and Ethical Implications for the Public Service", *Proceedings of the Public Service Management Conference* (Wellington: 1993) 23–35.
5. Robert Joseph, "Constitutional Provisions for Pluralism, Biculturalism, and Multiculturalism in Canada and New Zealand: Perspectives from the Quebecois, First Nations and Māori", Draft (Laws and Institutions for Aotearoa/New Zealand Te Matahauariki Research Institute: Wellington, 2000) 7.
6. Awhina Tamarapa, "Museum Kaitiaki: Māori Perspectives on the Presentation and Management of Māori Treasures and Relationships with Museums", *Curatorship: Indigenous Perspectives in Post-Colonial Societies: Proceedings* (Ottawa: Canadian Museum of Civilisation with the Commonwealth Association of Museums and the University of Victoria, 1996) 160–169.
7. Paul Tapsell, "The Flight of Parerautututu: An Investigation of Taonga from a Tribal Perspective", *Journal of the Polynesian Society*, Volume 106, Number 4 (1997): 323–373.
8. For instance, *taonga* possess *mana* (status, authority, power, energy), *tapu* (sacred, prohibition, indication of the presence of ancestors) and *mauri* (lifeforce). See Hirino Moko Mead, "The Nature of Taonga", *Taonga Māori Conference* (Wellington: Cultural Conservation Advisory Committee, Department of Internal Affairs, 1990) 164–169;

Sidney Moko Mead, "The Ebb and Flow of Mana Māori and the Changing Context of Māori Art", *Te Māori. Māori Art from New Zealand collections*, ed., S. Mead, (Auckland: Heinemann, 1984) 20–36; Tapsell, pp. 323–373.

9. See Moko Mead, pp. 164–169; Amiria Henare, "Nga Aho Tipuna: Māori Cloaks from New Zealand", *Clothing as Material Culture*, eds., S. Kuchler and D. Miller, (Oxford, New York: Berg, 2005b) 121–138; Tapsell, pp. 323–373.

10. Ngahuia Te Awekotuku, "The Role of Museums in Interpreting Culture: He Whare Taonga, He Whare Korero Are Museums Really Necessary?" *Art Galleries and Museums Association of New Zealand Journal*, Volume 19, Number 2 (1988): 36–37; Tapsell, pp. 323–373.

11. Paul Spoonley, "Becoming Pakeha: Majority Group Identity in a Globalizing World", *Sovereignty Under Siege? Globalization and New Zealand*, eds., R. Renwick and C. Rudd (Aldershot and Burlington: Athenaeum Press Ltd, 2005) 97–110.

12. Pastoral leases are a form of Crown land tenure unique to Australia, created by the British Colonial Office. They continue to be governed by Statute (http://www.nlc.org.au/html/land_native_wik.html Accessed 4 May 2008).

13. Ken Gelder and Jane Jacobs, *Uncanny Australia: Sacredness and Identity in a Postcolonial Nation* (Melbourne: Melbourne University Press, 1998) 136.

14. Gelder and Jacobs, p. 136.

15. Arapata Hakiwai, "Museums as Guardians of our National Treasures", *Art Galleries and Museums Association of New Zealand Journal*, Volume 19, Number 2 (1988): 37–39.

16. Jeavons Baillie and Lyndsay Knowles, "Conservation: The Case for Collective Action", *Art Galleries and Museums Association of New Zealand Journal* Volume 19, Number 3 (1988): 19–22.

17. Article 4, NZCCM Code of Ethics.

18. Auckland War Memorial Museum, National Museum of New Zealand Te Papa Tongarewa (Wellington), Canterbury Museum (Christchurch) and Otago Museum (Dunedin).

19. Gerard O'Regan, *Bicultural Developments in Museums of Aotearoa: What is the Current Status?* (Wellington: Museum of New Zealand Te Papa Tongarewa, National Services Museums Association of Aotearoa New Zealand, 1997).

20. Awhina Tamarapa, "Museum Kaitiaki: Māori Perspectives on the Presentation and Management of Māori Treasures and Relationships with Museums", *Curatorship: Indigenous Perspectives in Post-Colonial Societies: Proceedings*. (Ottawa: Canadian Museum of Civilisation with the Commonwealth Association of Museums and the University of Victoria, 1996) 160–169.

21. O'Regan.

22. David Butts, "Māori and Museums: The Politics of Indigenous Recognition", Unpublished PhD Thesis, Department of Museum Studies Massey University (Palmerston North: 2003) 11.

23. Butts.

24. Council of Australian Museum Association Inc, *Previous Possessions, New Obligations Policies for Museums in Australia and Aboriginal and Torres Strait Islander Peoples* (Melbourne: Council of Australian Museum Associations Inc, 1993). This policy was revised in 2005 with the publication of *Continuous Cultures, Ongoing Responsibilities*.

25. Council of Australian Museum Association Inc, 9.
26. Heritage Collections Committee of the Cultural Ministers Council *The National Conservation and Preservation Policy for Movable Cultural Heritage* (http://www.nla.gov.au/preserve/cult.html [Accessed 29 Sept 2007]).
27. Marcelle Scott, "Conservation in Australia – 1999", Opening Plenary Address, ICOM – CC Triennial Meeting, Lyon, France (1999).
28. Marcelle Scott, "AICCM Statement of Apology", *AICCM Newsletter* (2000): 1.
29. At time of writing, the AICCM local membership numbers 481, while the NZCCM has 148 members. However, the membership numbers of both organizations includes non-conservators in allied professions, such as librarians, archivists, framers, etc. All members, conservators and otherwise, were invited to participate.
30. Frank Matero, "Ethics and Policy in conservation", *Conservation, The GCI Newsletter*, Volume 15, Number 1: (2000): 5–9. Only 11% of those who were members of both organizations expressed an interest in completing a survey.
31. Alison Wain, "To Infinity and Beyond! A Little Light Crystal Ball and Navel Gazing for the Conservation Profession", *AICCM Newsletter No 90* (2004): 3–6.
32. Group based on shared ancestry.

18

Conservation, Access and Use in a Museum of Living Cultures

Marian A. Kaminitz and W. Richard West, Jr with contributions from Jim Enote, Curtis Quam, and Eileen Yatsattie

Introduction

In 1989 the National Museum of the American Indian (NMAI) was a fresh idea waiting to be built and to embody 'the first national museum dedicated to the preservation, study, and exhibition of the life, languages, literature, history, and arts of Native Americans.' As the sixteenth museum of the Smithsonian Institution, the NMAI 'works in collaboration with the Native peoples of the Western Hemisphere to protect and foster their cultures by reaffirming traditions and beliefs, encouraging contemporary artistic expression, and empowering the Indian voice.'[1]

The NMAI would consist of three buildings guided by input from Native Americans throughout the western hemisphere.[2] W. Richard West, the museum's founding director, inspired and guided the force behind both the construction of the three tangible edifices that comprise the museum's physical presence,[3] and the more intangible manifestation of an ideology, philosophy, and ethical and moral stance that has permeated everything from consultations with indigenous communities to the day-to-day operations in all areas of the museum's programming, including outreach to Native communities that the NMAI has termed 'the fourth museum.'

> From his arrival in 1990, West understood that this was an undertaking of great social and political significance for the nation, and he took steps to ensure that its evolution faithfully reflected the stake that so many Americans had in it. By pushing back the construction schedule two years, he allowed museum planners to consult with contemporary native communities throughout the Americas. From 1991 to 1993, the museum hosted two dozen consultations, attended by hundreds of people. The result of their involvement profoundly influenced both the design of the museums and the programs they carry out.

Native peoples did not wish to be seen as 'cultural relics' but as 'peoples and cultures with a deep past who are very much alive today,' says West. They also wanted 'the opportunity to speak directly to audiences through the museum's public programs, presentations, and exhibits; to articulate in their own voices and through their own eyes the meaning of the objects in our collections and their import in native art, culture, and history.'[4]

The Mission of the National Museum of the American Indian

The Mission Statement that guided the NMAI in its formative and developmental years, 1989–2005, put forth the following:

> The mission of the National Museum of the American Indian is to recognize and affirm to Native communities and the non-Native public the historical and contemporary cultures and cultural achievements of the Native peoples of the Western Hemisphere by advancing, in consultation, collaboration, and co-operation with Native peoples, knowledge and understanding of Native cultures including art, history and language and by recognizing the Museum's special responsibility, through innovative public programming, research, and collections, to protect, support, and enhance the development, maintenance, and perpetuation of Native cultures and communities.[5]

The Mission contained a number of strategic directives and values that not only guided the institution generally, but also had particular pertinence to the programmatic direction and shape of conservation.

The Mission affirms the fundamental nature of the NMAI as an international institution of living Native cultures of the Americas. It speaks directly to the role of NMAI as a hemispheric cultural center rather than a 'museum' in classic and conventional terms of the nineteenth and early twentieth centuries. The NMAI is a place of Native peoples and communities as well as their vast collections of cultural patrimony and is about the complex and numerous connections among them. Furthermore, the institution is constructed museologically from the 'bottom up' and the 'outside in' with respect to Native communities, thus making it a museum where Native peoples and communities have genuine intellectual and cultural sway, and where collaboration and mutual participation characterize relationships between the NMAI and its Native constituents.

The NMAI and contemporary native peoples and cultures

The NMAI has numbers of laudable objectives, but perhaps none more important than the Mission value of demonstrating the contemporaneousness of the Native peoples and cultures of the Americas. A principal legacy of the nineteenth and twentieth centuries was the perception, which ultimately became an axiom of popular culture, that Native peoples, unfortunate victims of military and cultural conquest, were permanently off the stage of history. Coupled with the academic norms of the day, this conception made it far easier to view Native peoples and their interpretation as a matter of museum ethnography alone – cultural reclamation projects defined by reference to material culture.

Twenty-first century demographics belie completely the notion of Native peoples of the Americas as dead or dying. Notwithstanding demographic catastrophes that occurred over almost five centuries, Native populations throughout the Americas are recovering. The Native population in the United States has surpassed 2,000,000, the peoples of Canada's First Nations number over 1,000,000, and the indigenous populations of Latin America probably approach 40,000,000. This long-time resilience is compounded by the twentieth-century rebirth of a profound cultural commitment on the part of Native communities to maintain and sustain culture and cultural practice into a twenty-first century future. The NMAI is committed to convincing every visitor who walks through its doors on the National Mall or into its George Gustav Heye Center in New York that Native peoples, cultures, and communities – as well as their patrimony – exist across this broad time continuum.

Native authority at the NMAI

A second Mission value is the authority that is accorded Native peoples with respect to their representation programmatically at the NMAI. Historically, interpretive 'authority' concerning them has been held in third-party hands of anthropologists, archeologists, historians, art historians, or others grounded in non-Native learning disciplines and systems. Without at all questioning the legitimacy of that scholarship and those perspectives, the NMAI has used a fundamentally different methodology based upon the first-person voices of Native peoples themselves in interpretation and representation.

This approach permeates the entirety of the NMAI programmatically and is the basis for all of its major policies relating to exhibitions, public programs, repatriation, research, publications, conservation, and others. The methodology has been most evident, because it is the most traditional and publicly acknowledged museum medium, in exhibitions. From its beginnings at the George Gustav Heye Center in New York with the exhibits, *Creation's Journey*, *All Roads Are Good*, and *This Path We Travel*, to the later permanent installations on the National Mall, *Our Universes*, *Our Peoples*, and *Our Lives*, as well as the changing exhibitions, *Listening to Our Ancestors* and *Identity by Design*, the NMAI consistently has invoked, on an unfiltered basis, the cultural expertise and authenticity of Native peoples in articulating their own cultural and historical viewpoints and knowledge in its exhibition galleries.

The NMAI and sustaining contemporary native cultures

The vital third leg of Mission values involves responsibilities of the NMAI to contemporary Native cultures and communities. No aspect of the institution is more essential to its standing as an international institution of living cultures than this characteristic. The NMAI commitment is best symbolized organizationally by the museum's Community and Constituent Services Group, which is responsible for overseeing and maintaining relationships with contemporary Native communities and for promoting their maintenance of culture into the future. This philosophical and operational tenet, however, infuses all the work of the NMAI, stretching beyond the Community and Constituent Services Group and also driving the operation of other organizational groups including Museum Assets and Operations, the group in which collections and conservation departments are seated.

More specifically, this commitment to the maintaining and sustaining of culture in contemporary Native communities has had dramatic impact on the work and practice of the NMAI's conservation department. The range of questions and issues that arise in placing conservation practice in the context of the foregoing discussion are many and complex and, in summary form, can be described here as follows. First, the material culture that sits in the vast collections of the NMAI is vital – indeed, probably central – to sustaining culture in contemporary Native America. The collections embody, carry, and, most importantly, perpetuate culture and its practice because they encompass fully those powerful essences of meaning as tangible evidence. Preservation of these tangible collections is important because they are pivotal source points for those in contemporary Native communities who wish to maintain, and prosper in, cultural cohesion.

Second, the NMAI's sharing of authority with respect to the interpretation and representation of Native culture has direct impact in the conservation area. Not only does the institution recognize, from the standpoint of classic museum object conservation, that it has much to learn from the Native communities that, through time, themselves had a profound commitment, in many instances, to the physical protection of objects; but for the NMAI, 'conservation' also takes on cultural rather than only physical dimensions. Objects in the collections of the NMAI are not viewed as 'inanimate' by their makers; they are profoundly and perpetually 'alive' – and must be so treated. This dimension demands that cultural protocols, tangible and intangible, surround the objects and their care – not historically the domain of museum conservation departments.

Finally, the two previous points mean that the NMAI must focus on the creation of access to the material in ways almost counter-intuitive to classic conservation, where the impulse was to protect, and thus to limit, physical access to museum objects. Instead, the NMAI has committed itself, in a variety of ways, to working with Native communities throughout the Americas, to create, in a spirit of collaboration and mutuality, access, physical and virtual, to its vast material collections.

Conservation and NMAI's mission

The conservation department is an active participant in the manifestation of the NMAI mission. In 1991, as conservation began to develop along with the new museum, the fresh, inspired ideas of the NMAI's programming began to blend personal ethics with a holistic humanistic approach to work. This programming was buttressed by, and relied in large part on, the Museum's mission statement, which stressed 'consultation, collaboration, and co-operation with Native Peoples . . . to protect, support, and enhance the development, maintenance, and perpetuation of Native cultures and communities.'[6] Acknowledging the diversity of indigenous cultures and continuity of cultural knowledge, staff began integrating and incorporating into daily processes Native methodologies for the handling, documentation, care, and presentation of the collections. As this occurred, the staff's professional conservation ethics in the department began to undergo a process of revision and adaptation as a response to working with living indigenous cultures of the Western Hemisphere. We began to realize that while the tangible collection objects were the visible evidence of culture, they were just one part of what makes up the culture. Songs, stories, lineages to ancestors, cultural constructs, acknowledgement of the collection items as living, and much more, were all linked intangibly to the physical items we conserved. These linkages were embodied in the people and descendants of those who made the items in the collection.

While the NMAI's mission statement provided the foundation for the Museum to build and organize its programs, in conservation it informed the way we undertook conservation treatments, who advised us on appropriate collection storage components and methods, how the conservation department was staffed with employees, interns, fellows and volunteers, and how we began to share conservation decisions by incorporating Native voice, values, and use into our conservation and preservation environments. It also pressed the conservation staff to question conservation academic ideas and assumptions as we assimilated NMAI mission values to allow the indigenous cultural owners of the collection a true voice. We tore down and built anew some learned behaviors and biases having to do with process and control; others we had to let go while still maintaining professional conservation ethics and standards alongside the ethics of working in a living museum.

Since the early 1980s, conservators have increasingly worked with Native, indigenous and local communities to determine conservation treatments for museum collections and public monuments.[7] This type of project planning has expanded to become a more commonplace and accepted route for conservation treatments. In numerous instances, indigenous representatives knowledgeable in cultural materials and technologies have partnered with NMAI conservators to undertake conservation treatments and/or care for museum collections in a culturally appropriate manner. During these consultative partnerships with indigenous representatives, we learned, were told, understood, allowed, practiced, and assimilated new methods and broadened our conservation processes. Pros and cons and practicalities learned from these experiences have been discussed in several previous NMAI conservation publications and will not be reiterated here.[8] Processes are always evolving and no two experiences are alike. Negotiating appropriate methods for conservation treatments occurs through respectful exchange and understanding that cultural sensitivities may override the need for treatment by a staff conservator. What follows is an example of NMAI outreach in which conservation supports community initiatives.

Hawikku pottery comes home to Zuni, 2001 – the present

June 22–23, 2001: the NMAI conservation department was involved in a community consultation with Zuni tribal members, to guide and direct conservation treatments in a culturally sensitive and appropriate way as items were prepared for a long-term loan to the Zuni Pueblo in New Mexico. This project, many years in the making, would result in over seventy pottery items from the Hawikku excavations of 1917–1924[9] being sent back to the Zuni community for an exhibition and as a handling collection at the A:shiwi A:wan Museum and Heritage Center (AAMHC).[10] Zuni tribal members, Eileen Yatsattie, a well-known and respected traditional potter, and then Lt Governor Barton Martza, who had been working towards this loan for over 15 years, came to NMAI along with then AAMHC director Tom Kennedy, to finalize selections of the Hawikku pots and consult with conservators.

NMAI conservation consultation process

The NMAI presence in the consultation was purposefully kept small, with one curator and three conservators, to assure the Zuni would be comfortable as the directors and owners of the consultation process. Most Western cultures have very different styles of approaching the handing down of cultural knowledge when compared to those of indigenous communities where particular knowledge is often restricted to specific individuals. Learning methods are also different. Asking questions directly and outright is often considered impolite; rather, students learn by watching and listening carefully.

The importance of ownership of the consultation process by community representatives should not be underestimated. It allows the cultural representatives the opportunity to state what they want and how to do it, it builds a partnership environment, and it allows a secure trust to grow with the knowledge that their interpretation of their own cultural material will be what is presented in the final interpretive process (exhibition, loan, conservation treatment, etc.) instead of the museum's interpretation. It also ensures first-person voice.

The two-day consultation gave us time to get comfortable with one another. Eileen talked about fabrication methods, stylistic approaches, construction techniques, surface finishing, and design imagery. When Eileen looked at the pots, a lot of emotions arose. She related to the pottery as ancestors of her people. She smelled each pot to see if it had held water from Hawikku. She examined the clay body to determine if it was the local clay or brought in from outside the community. She handled each pot to see how it was made and how good the potter was in her execution of the vessel. She looked at the imagery painted on the surfaces and interpreted it for us.

We conservators showed Eileen how we would surface clean the pots with a soot removal sponge to lift off more than 80 years of sooty particles from storage at the Museum's old facility in the Bronx.[11] She said this was appropriate to do and that we could remove old adhesive residue from the original reconstructions carried out by excavators and restorers in the 1920s. She also agreed to small amounts of local consolidation in areas of flaking surface. When we discussed whether to consolidate or desalinate a whole vessel that was badly damaged from soluble salts in the Hawikku environment, Eileen did not wish us to undertake this type of treatment. From her perspective, this would remove the use history of the pot. To gauge the conditions and better understand the environment in Zuni where the items would be exhibited, we spoke with Tom Kennedy about the museum building and we researched annual climate conditions for the area. While there were wet and dry periods during the year, we felt that we should follow the community's requests and monitor conditions rather than push to treat the pot more invasively. Also, we were not to ship the pots in plastic bags because the pottery is considered to be living ancestors of the Zuni.

Return to Zuni

In late July 2001, the pottery arrived in Zuni. Friday, July 27th's edition of *The Shiwi Messenger* (the local Zuni newspaper) captured the local importance of this event in W. Makee Jr's cover article, 'Hawikku Pottery comes Home to Zuni.'[12]

> Dark rain clouds loomed overhead as a small group of Zuni community members gathered outside the Zuni Christian Reform Church. Despite the gloomy atmosphere, there was a sense of happiness and enthusiasm in the air. The reason was simple.
>
> The *A:shiwi* (Zuni people) believe that our ancestors sometimes come to check on how we are doing and to bestow blessings upon our people, and

they usually come in the form of rain. Needless to say, the Zunis gathered outside the parking lot early Thursday morning had plenty of reason to believe that our ancestors had arrived to welcome back a small piece of our illustrious history.

A little over 80 years ago, in 1917, excavations started at the ancestral site of Hawikku. Some Zuni elders recalled that as children, they had seen wagonloads of boxes leaving Zuni. Many of them wondered what was in those boxes and where they were going. Back then, our elders had no idea that they were witnessing history. By the end of 1924, approximately 20,000 artifacts had been taken, almost certainly never to be seen again in Zuni.

Almost eighty years later, history is being made again. The Pueblo of Zuni, A:shiwi A:wan Museum and Heritage Center, and the Smithsonian Institution, National Museum of the American Indian (NMAI) are working together to bring some of the items that were in those boxes back to Zuni, in essence, bringing them back home.

. . . Today, people like Eileen Yatsattie are . . . trying to piece together the mystery that our ancestors left behind. When the first 'kiva'[13] was opened, a sense of awe filled the room as the centuries-old pottery vessels contained within seemed to tell us the story of its life, a life now filled with happiness now that the spirit of its creator had finally returned home. We could finally say, 'Ancestors of *Hawikku,* welcome home!'

Marian Kaminitz's 2007 visit to Zuni

The physical conservation of the Hawikku items returned to the Zuni community is only part of the story of the importance of cultural involvement in the treatment process; the real commitment to continuance of culture occurs within the community. In an effort to find out if the return of the Hawikku items had assisted in this, six years later (from 30 July 2007 through 2 August 2007) I visited Zuni. During the visit, I condition-checked the Hawikku items for NMAI's loan records – it should be noted that none of the pots showed any damage, including any from soluble salt activity. I also discussed the impact of having the ancestral materials back at Zuni, talked about possible future loans, and obtained permission to use information from those discussions and the loan experience in this publication. To respect cultural protocols, I asked for permission and the Zuni Tribal Council agreed to the overall idea and asked me to work directly with AAMHC staff. AAMHC Executive Director James (Jim) Enote and Museum Technician Curtis Quam were directly involved in my visit and discussed the loan with me, as did

Eileen Yatsattie. To respect first-person voice, the following excerpts from those conversations have not only been approved by the speaker, but are also left intact so that the reader may 'hear' the speaker's perspective without it being restated or unintentionally misinterpreted.

1 August 2007 interview with Eileen Yatsattie (EY)

MK: Why is it important that the material is at Zuni?

EY: Zuni pottery [is] all over the world, but you hardly see it back home. I'm very selective where I market my pottery. I prefer to market it to my people so that it will stay in Zuni. Seeing the Hawikku collection got me . . . to do more utilitarian pottery so that it can stay here in Zuni so that the people can use it. . . . [Seeing] all the different uses, I think that is the way [the experience with the Hawikku pottery] really benefited me . . . I like to continue what my grandmother and great-grandmother did.

About three families are doing traditional pottery using traditional materials. I was hoping by bringing this material home . . . [I could] show my people, the Zuni people, what's out there at museums. Some of these pots are very unusual that our ancestors made a long time ago . . .

It made a big impact on me also. I'm teaching my grandkids – nephews and nieces to stay within the traditions and to continue what our people did. Most of the traditional arts are dying down. . . .

. . . The idea is to have the pieces as close to home as possible. It gives young kids the ability to question about their history; it gives them opportunity to learn. [It] captures [their] attention, making them more appreciative of their culture and religion – appreciate . . . being Zuni – teaching them history and family values and morals. By teaching the kids family values, it tends to keep them out of trouble.

. . . It helps the community cause it has the chance to turn the kids around.

2 August 2007 interview with Jim Enote (JE) and Curtis Quam (CQ)

MK: What does it mean to you that the Hawikku pieces came back to Zuni?

JE: It's really wonderful that they are here because they belong to Zuni . . . It's allowed Zuni to reflect on the idea of repatriation of how we look at old things from our past and whether some things are appropriate or not to be here. It's allowed us to think more critically about the meaning of having excavated objects back at Zuni. It has opened peoples' eyes to think in many different ways about what a museum should and should not have in its collection.

CQ: I think it's good to have the artifacts back. It makes me think more about things that have happened in the past – wondering about how things were made and used. Now it's artwork but before it was utilitarian. I think about how things have changed since those artifacts were used. The designs bring up a lot of questions in my mind. When I get some of those questions answered or when they become more apparent, I feel closer and have a stronger connection to my ancestors, culture and people.

MK: Why is it important?

JE: It has provoked thoughts about what is appropriate for viewing and maybe also what are Zunis interested in . . . see[ing] in the Heritage Center/ museum? When I see that there's ceramics in there, there's one type of Zunis that like to see that sort of thing and there are other kinds of Zunis that don't want to see it.

MK: Is it an instance of making something they relate to?

JE: Yes. It is a first major exhibit for the AAMHC so, it's always something we can learn from, make better and we shouldn't think this is the only exhibit that will ever be here. . . . I . . . feel that the process of creating the exhibit is . . . part of that whole experience. We . . . bring advisors together and give them the opportunity to . . . create an exhibit. That exhibit design process engages a large group of Zunis . . . designers, laborers . . . there are a lot of people actually behind something like this.

CQ: Watching the youth come into the museum and go through the exhibit – how it's laid out, I think it's more appropriate for older people . . . When I see families come in, it's more of a learning place. If they [kids] are with their parents, they have a different demeanor – they listen more to what's going on. It's important to present history and that's what the exhibit is doing. It brings up wondering how things have changed and thinking about things like how the language was then.

JE: . . . I think it is important that these pieces bring something physical to 'long time ago.' It wasn't just 'long time ago' people, this is something they made, it was really used. . . . I think it is really important that at *this* time the young generation should see something that connects us to our past. My grandparents didn't have to see it – they were just told it and believed it.

MK: Do you think it helps people know more of who they are by knowing about their past?

JE: I think it does. When Curtis talks with school groups . . . they hear about the past but they actually get to see something physical. . . . If you watch kids come through, they don't stop to read the text. They focus on the objects but for a very short time.

And then maybe there's another group of Zunis that I wonder if they think . . . 'Museums are about pots and bones.' Maybe some Zunis want that and maybe some want to stay away from a place like that. Maybe some don't come because they think it's a pots and bones place.

Some Zunis say: 'That's where white people go.' They think it's not really for them because maybe non-Indians are more interested in pots and bones and they are more attracted to that and maybe Zunis say, 'If they are interested in that, I won't be.'

. . . The center is more than just a museum.[14] It is a place to explore different modes of learning. Essentially we are exploring how we learn across knowledge systems: Zuni and Western. That's what we are trying to advance here at the center.

Conclusion

As part of the Smithsonian Institution, NMAI's collections are legally considered to be owned by the citizens of the United States. However, in a very practical and functional way, NMAI staff consider themselves the stewards of the collection, which is culturally owned by the indigenous constituents and source/descendant communities of the items in the museum. Our partnership methodology helps promote cultural self-determination and continuance both tangibly and intangibly, while it also challenges us as ethnographic conservators to re-address our ethics and re-define ourselves as professional conservators. By working in the Museum's conservation department with community involvement as well as in the communities themselves, we have gained personal experiences and developed skills that allow and guide us to evolve our conservation methodologies.

'Intangible heritage' Richard Kurin noted, 'is by definition living, vital and embedded in ongoing social relationships.' 'In museums, objects become part of collections and reside under the roof and authority of the museum. With intangible cultural heritage, the traditions exist outside the museum, in the community. They reside under the authority of the people who practice them . . .' and 'they must have the major role in defining their own intangible cultural heritage and how it is documented, preserved, recognized, presented, transmitted, and legally protected.' 'By working closely and co-operatively with the relevant communities,' museums and national organizations can be key partners in the 'presentation, preservation, protection and transmission' of intangible and tangible cultural heritage.[15]

Honoring cultural integrity and providing for the care, preservation, expansion, and use of NMAI collections is a process that includes supporting dynamic access to and care of the collections by Native communities. The outcome, a continually evolving conservation and collections care methodology responsive to living indigenous cultures and traditional values, helps to assure the continuance of cultural heritage and is filled with discovery, creativity and growth.

Acknowledgements

Grateful thanks are given to Jim Enote and Curtis Quam for the time they spent reviewing the Hawikku materials at the AAMHC and for their wisdom and contribution to this chapter. Thanks as well to the AAMHC for permission to reuse the quote from the Shiwi Messenger newspaper.

Eileen Yatsattie's knowledge of pottery enlightened and guided NMAI's approach to conserving over 70 pottery vessels for the exhibit. We are indebted to her willingness to share her understanding of traditional Zuni pottery.

Consulting with communities in conservation processes would not be possible without generous support from the Andrew W. Mellon Foundation.

Notes

1. NMAI website accessed 10 March 2008 (http://www.nmai.si.edu/subpage.cfm?subpage=visitor&second=about&third=about).
2. United States Public Law 101-185 – Nov. 28, 1989. National Museum of the American Indian Act. 103 STAT. 1136–1347, accessed 10 March 2008 (http://www.repatriationfoundation.org/%20PDF%20Files-%20MTC%20MainBook/Appendix%20F,G.pdf or http://thomas.loc.gov/cgi-bin/query/D?c101:1:./temp/~c101e8PDQp::).
3. The National Museum of the American Indian has three sites. The George Gustav Heye Center, a public program and exhibition site in the Alexander Hamilton United States Customs House is in New York City and opened October 1994. The Cultural Resources Center in Suitland, Maryland, 8 miles southeast of Washington DC, houses the collection and includes conservation staff offices and labs. The NMAI Mall Museum, a public program and exhibition site, opened 21 September 2004 in Washington, DC. 10 March 2008 (http://www.nmai.si.edu/subpage.cfm?subpage = visitor).
4. Mark Jacobs, "W. Richard West: Museum Director – Achieving a Dynamic Balance," *eJournal USA: The United States in 2005: Who We Are Today (December 2004)*, 10 March 2008 (http://usinfo.state.gov/journals/itsv/1204/ijse/west.htm).
5. The 2005 revised Mission statement is: 'The National Museum of the American Indian (NMAI) is committed to advancing knowledge and understanding of the Native cultures of the Western Hemisphere, past, present, and future, through partnership with Native people and others. The Museum works to support the continuance of culture, traditional values, and transitions in contemporary Native life.' National Museum of the American Indian website, 10 March 2008 (http://www.nmai.si.edu/subpage.cfm?subpage=press&second=aboutnmai).
6. NMAI Mission Statement cited in "National Museum of the American Indian's Collection Policy," Agnes Tabah, Native American Collections and Repatriation: Technical Information Service's Forum – Occasional Papers on Museum Issues and Standards (Washington, DC: American Association of Museums, 1993) 166.
7. Miriam Clavir, "Reflections on Changes in Museums and the Conservation of Collections from Indigenous Peoples," *Journal of the American Institute for Conservation*,

Volume 35, Number 2 (1996): 99–107. (http://aic.stanford.edu/jaic/articles/jaic35-02-002.html); Monika Harter, Marian Kaminitz, Kelly McHugh, Melinda McPeek, Leanne Simpson, and Gary Raven, "Appropriate knowledge, appropriate action: a request for the treatment of eleven Midewiwin birch bark scrolls," AIC Abstracts of the *American Institute for Conservation 29th Annual Meeting, Dallas* (Washington, DC: AIC, 2001) 2–3; Susan Heald and Kathleen Ash-Milby, "Woven by the grandmothers: Twenty-four blankets travel to the Navajo Nation," *Journal of the American Institute for Conservation*, Volume 36, Number 3 (1998): 334–345; (http://aic.stanford.edu/jaic/articles/jaic37-03-007_indx.html); ed., Denyse Montegut, Preprints: Caring for American Indian Cultural Materials: Policies and Practices – a Two-Day Symposium, October 19–20, 1996 (New York: Fashion Institute of Technology, Graduate Division, 1996); Nancy Odegaard, "Artists' Intent: Material Culture Studies and Conservation," *Journal of the American Institute for Conservation*, Volume 34, Number 3 (1995): 187–193 (http://aic.stanford.edu/jaic/articles/jaic34-03-003_indx.html).

8. NMAI conservation staff publications about experiences with indigenous communities are listed at: (http://www.nmai.si.edu/subpage.cfm?subpage=collections&second = conserv&third=staff_bib).

9. The NMAI collection items loaned to the Zuni, were chosen from the Hendricks-Hodge 1917–1923 archaeological excavations of the ancient Zuni city of Hawikuh for the exhibition entitled *Hawikku: Echoes from Our Past*. The exhibit explores, from the Zuni perspective, the history of the relationship between Zuni and Euroamerican culture from the emergence of the Zuni to the present.

10. A:shiwi is the Zuni's name for themselves. The A:shiwi A:wan Museum and Heritage Center is located in Zuni Pueblo, Zuni, New Mexico.

11. The Museum of the American Indian – Heye Foundation's exhibition facility was located at 155th and Broadway, NYC, and was the precursor to the NMAI. The museum's storage site in the Bronx, NY had particles from leaded gasoline vehicle emissions and soot from surrounding buildings' coal fired furnaces infiltrate and accumulate for 80+ years onto the open shelving units where the Hawikku pots were stored.

12. Wells Mahkee, Jr., "Hawikku Pottery Comes Home to Zuni," *The Shiwi Messenger*, Friday, July 27, 2001 issue, Volume 7, Issue 15 (Zuni, NM: A:shiwi A:wan Museum and Heritage Center, 2001) 1.

13. The reusable commercial corrugated plastic shipping containers that the Hawikku pots were sent in to Zuni are trademarked under the name 'Kiva'. In Zuni and other Puebloan cultures, a Kiva is a religious meeting house.

14. For an in-depth study of how the Zuni people made the AAMHC a culturally relevant public institution to help them maintain their heritage for future generations, see Gwyneira Isaac, *Mediating Knowledges: Origins of a Zuni Tribal Museum* (Tucson, AZ: University of Arizona Press, 2007).

15. Richard Kurin, "Museums and Intangible Heritage: Culture Dead or Alive?" *Museums and Intangible Heritage: ICOM News*, Volume 57, Number 4 (2004): 7.

19

The Challenge of Installation Art

Glenn Wharton and Harvey Molotch

Introduction

The rise of installation art challenges principles developed for conserving traditional media. The conventional canon to honor the 'authentic' object, already under some stress, becomes especially problematic when dealing with art whose meaning and materiality cannot be fixed.[1] In this chapter we outline the challenges that arise in conserving such ephemeral and contingent works. We suggest changes – some already underway – through which conservators may respond.

Contemporary installation art draws inspiration from earlier projects not meant for collecting institutions at all. Along with the closely related genre of performance art, installations of the 1960s were anti-establishment in general, and anti-museum in particular. Their temporary quality was often part of their point. Included in these immediate precedents were *Happenings* created by Allan Kaprow, Fluxus projects, and other Dada-inspired events. When such works yielded artifacts at all, they were more akin to happenstance props than objects to be conserved. Performances were one-offs, mounted in public spaces for public interaction – activist interventions absurdist in spirit or aimed against establishment politics and on behalf of counter-cultural ideals.

As gallery owners, collectors, and museum curators took note, they initiated means to display, celebrate and eventually 'own' such projects or at least those they inspired. In so doing, they pressed these works into the standard modes of operation, not only of display but also of registration, storage, and conservation. Some artists responded by shaping their works to better operate within the boundaries of the art world, while the institutions stretched to respond to the artworks' exigencies. The process continues along a two-way street of adjustment.

Although sometimes retaining an oppositional frisson, installation art now ranges to the benignly serene. In physical content, some works utilize simple materials like a roomful of dirt (Walter de Maria, *The New York Earth Room*), while others involve performance and computer-generated displays. Joan Jonas combines performance with video and related artifacts in her installations such as *Revolted by the Thought of Known Places*, 1994 (Figure 19.1). Whatever else it may be, re-presenting such work cannot mean finding out how to replicate the exact nature of an authentic original. In a real sense, there is no hallowed original, the first showing having been a result – in varying degrees – of happenstance, imposed limits, or events of a historic moment that have passed into oblivion. Instead, the capacity to perpetuate the art, in some way or another, depends on capacities and conditions in the *present* moment and not just on those in the past. The conservator works with others, and a series of physical, institutional and technical contingencies – as we now discuss – to arrive at decisions about how the artwork can continue.

Figure 19.1 *Joan Jonas,* Revolted by the Thought of Known Places . . . Sweeney Astray, *1994. Performance view. Westergasfabriek, Amsterdam. © Joan Jonas. Reproduced with permission from the artist and Yvon Lambert New York, Paris.*

I. Physical context

Physical context always helps define a work of art, but this becomes more radically the case with installations that use environment to structure viewer experience.[2] Room dimensions, windows, doors, and interior boundaries can be intrinsic to aesthetic goals, along with sight, sound, and aroma. Dan Flavin's arrangements of fluorescent fixtures exist as sculptural components within a larger expression of color and light intensity. In how large of a room can a Flavin be exhibited while retaining the desired luminance?

Installations often lack clear boundaries. The term for a sub-genre of installation art, 'Scatter Art,' signals the problem. Certainly at the time of reinstallation, staff may lack knowledge of edges of the work or just how patrons are meant to walk around or within them. They place stanchions, platforms, and floor lines to guide viewers and define what's 'in' and what's 'out' – an aesthetically and functionally delicate maneuver. Any reinstallation needs to somehow be appropriate to the relationship between the work itself and the other elements of the room, including bodies of visitors. Knowing the artist's tolerance for variation, or maintaining the same boundaries are aspects of conserving the work.

Replicating spatial characteristics of earlier installation environments raises dilemmas.[3] Adjacent spaces leading into the work may not be the same. The original room or building may no longer exist. Even if there is sufficient money to reproduce the old room within the walls of a newer space, this may lack the right 'feel.' A part of the original idea may have involved using the building's found qualities as an aspect of the work.

II. Artifact status

The amalgams that can make up a single installation – toys, food, commercial projectors, and so forth – force conservators to make different judgments about the value of the components. Although some prioritizing occurs in general conservation practice, installations force the issue of deciding where the effort should go. Elements within a given work thus take on varying *status*. Just as individuals differ in social standing, each element of an installation has a particular standing vis-à-vis every other element. At one end of the continuum, an element may be crucial; without it, the work loses all meaning. Daniel Buren created a kind of kaleidoscope spectacle by fitting the Guggenheim's famous spiral atrium with mirrored panels (Daniel Buren, *Around the Corner,* 2000–2005). Substituting plain glass (or non-reflective panels of some sort) in a subsequent installation would obviously be ridiculous. Mirror has high status. At the other end of the spectrum might be a rag whose fabric, shape, or soiling is incidental.

The shape and finish of a common screw may be insignificant in one project but critical in another. If judged as low status – simply a way to hold two elements together, for example – one could pull a replacement with similar thread count from the supply cabinet. Or find a different way of holding things together, with a hinge perhaps. With higher status placed on the original, conservation may require re-threading the existing screw or replacing dislodged metal with fill material – a time-consuming process. Low-status artifacts can also include behind the scenes projectors, media playback equipment, and replaceable parts such as light bulbs and projector filters. Artists may create their own replacement components or owners may purchase or fabricate them, sometimes tossing them after the exhibition closes.

Artifact status can also change. What was first considered low status might become high status, and vice versa. A single element can also have mixed status, high in one regard, but low in another. An example of mixed status is the candy used in *Untitled (USA Today)* 1990, a typical Felix Gonzalez-Torres 'spill' (Figure 19.2). The candy is low status in that people consume it and it is therefore disposable. Indeed, although the artist indicated a certain 'ideal' for his candy (in terms of weight) for example, he did not require that it be of a specific type. But it is crucial that there *be* candy, and conservation includes understanding which candy it should be.

Figure 19.2 *Felix Gonzalez-Torres, "Untitled" (USA Today) 1990. Museum of Modern Art. Gift of the Dannheisser Foundation. © The Felix Gonzalez-Torres Foundation. Reproduced courtesy of Andrea Rosen Gallery, New York.*

Error can occur. Without understanding the meaning assigned by the artist, subsequent caretakers can omit something considered important or pay too much attention to something that isn't. Such was the case when, somewhere along the line, somebody removed a two-inch thick layer of debris from an excised building segment intended for gallery display as part of a Gordon Matta-Clark 'anarchitecture' exhibit (Gordon Matta-Clark, *Splitting: Four Corners*, 1974). Ordinarily low status, dust and debris in this instance served as a sort of patina testifying to the building's destruction by the artist. It survived at least the first two exhibitions, but was erroneously cleaned before a subsequent display after the artist died.[4]

Another status difference among artifacts, particularly relevant for media installations, is whether the display equipment is *dedicated* or *non-dedicated*. Just what is the status of monitors, projectors, platforms, viewing screens, lights, sound equipment and hard disc drives? An institution may dedicate equipment for exclusive use of a particular work. Alternatively, the institution can pull equipment from the general appliance pool for a media exhibition, or for a public lecture for that matter. With non-dedicated equipment, media installations visually transform as available equipment changes over time. For the artist, aesthetic effect may be at stake; for the institution, there are financial, organizational, and real estate consequences of storing unique equipment for each media installation. Here again, the conservation outcome depends on how decision makers adjudicate the issue.

III. Physical transformations

Even with traditional art, conservators increasingly think of objects as having 'agency;' for instance, when color fading or breakage instigates preventive or interventive response.[5] But again, installation art elevates the issue because of the variety of materials, their differential rates of disintegration, and the fact that disintegration – of a certain sort – may be intrinsic to the work. Conservation may include quality control during installation to facilitate proper disintegration, or managing the environment to affect the rate of change and evacuate resulting fumes. Installation art shifts attention from the artifact towards sustaining the capacity to replicate *the event* as specified by the artist. Lee Mingwei and his team spend days creating an intricate rendition of Picasso's *Guernica* in colored sand on the gallery floor (Figure 19.3). While the artist is still in the gallery completing the project, a first visitor walks on it, followed by successive visitors who radically disfigure the Picasso-like image over the course of a day. Staff controls the specifics of visitor behavior, for example having parents stay with children who otherwise might destroy the painting 'too quickly,' as if they were in a sand box. An aesthetic tension comes from the near simultaneity of creation and destruction among the participants. That tension, together with the sand itself, is intrinsic and must be integral to future reinstallation.

Installations bring other ways of loss to the fore, including dilemmas that stem from the wear and tear of active *use*. Conservators, along with artists and others in the process, face the fact that some elements wear out through the fact of their exhibition. Installations may have internal moving parts or may be in interaction with gallery visitors who push buttons, bounce on platforms, or open doors and drawers. The decision to mount such installations has contradictory effects on longevity. On the one hand, the more an object operates, the shorter its physical life. On the other hand, installation *performs* the work in a way that increases its appreciation, including financially. This builds the affection needed to secure resources for future exhibition and conservation.[6] In contrast, installations relegated to storage grow cold as familiarity with them ebbs away.

The Challenge of Installation Art

Figure 19.3 *Lee MingWei,* Gernika in Sand. *Chicago Cultural Center 2007. Collection of Yeh Rong Jai Cultural & Art Foundation, Taiwan. Photo reproduced with permission from Anita Kan.*

As William Real points out, 'repeat performances' not only draw on conservation knowledge, but also bringing that knowledge back and refining it functions as a means of preservation in itself.[7]

Many artists include time as a medium; hence the phrase 'time-based media' is often used to characterize moving image art. Installations using such technologies are not only prone to wear and tear, but to format obsolescence. Videotape, optical discs, computer hard drives, and other information carriers all have limited life. Owners must regularly copy information from one carrier to another, or migrate it to new technologies (e.g. analog video to digital video). Computer-driven works may require expensive *emulation*, or re-writing software code enabling them to run within new computer environments.[8] A central conservation question is how to choose new formats as technology evolves. Migrating a film or analog video to digital may compromise (and certainly change) picture quality. There may be loss of important information when saving images to compressed formats, such as JPEG or MP3. Similar issues arise with media playback and display equipment. When an installation's slide projector wears out, should its images be digitized for projection on a digital projector? Is the work being preserved if there is no longer the 'click' sound that punctuates changing slides? Should an audio recording of the clicking be added in to compensate? To what extent is the medium itself the message?

IV. Available documentation

Another circumstance influencing the ongoing life of installation art is the accumulated documentation from production and acquisition to later research and re-installation. This accumulation of text, images, and instructions can sometimes communicate more about the work than its physical manifestation.[9] Production and acquisition can be an ambiguous process for a form of art that exists at the nexus of time, light, and motion.[10] In addition to the installation artifacts (if they exist), the collector gains a bundle of specifications and legal rights governing reproduction and display. A certificate of authenticity and other artist-produced documents may provide guidelines and step-by-step instructions. Available documentation at the time of re-installation varies radically across projects; prior installers may leave floor plans, diagrams, photographs, audio video recordings to assist future reinstallation. To the extent they exist, documentary traces from the past shape the institutional memory of what the work can be.[11] They better define it with each venue by establishing potential variability in spatial characteristics and artifact status.

Some artists go to great lengths to be clear. Bill Viola, for example, provides 'artist kits' or 'archival boxes' that tightly specify physical aspects of his video installations.[12] An artist can place restrictions on how and where works are shown, along with specifications from one installation locale to another. Among other exhibition criteria listed in his certificate for *Certificate of Glass (one and three)*, 1976, Joseph Kosuth states: 'It is the intention of Joseph Kosuth that this work be owned or exhibited exclusively in a Flemish speaking cultural/linguistic context. Fulfillment of this requirement is absolutely essential to the existence of the work (as art).'[13] Other artists, such as Nam June Paik, assign broad authority to curatorial interpretation in re-installing their works.[14]

But often exhibitors have little to go on. The Dia Art Foundation at times found itself with this predicament when installing works from its stored collection at the opening of its vast warehouse style museum in Beacon New York in 2003. For some installations, such as works by Robert Smithson, Dia purchased entirely new materials. Since he was no longer alive, staff resorted to the artist's writings, diagrams, and documentation from similar pieces.[15]

The conservator may find documentation in the form of database entries that characterize the work. As databases set up to catalog traditional art fail to capture all dimensions of installations, recent efforts create new vocabularies and metadata fields. The Variable Media Initiative[16] defines artwork in terms of medium-independent behaviors, like the presence or absence of human performance, and its connection to installed artifacts. Protocols entered in the catalog database provide ways to document what it means to install the work – like Felix Gonzalez-Torres' tolerance for variability in the shape and size of his piles of candy. Documenting that his guidelines for the candy spills are intended as 'ideals' to be used as reference and are open to interpretation is crucial to maintaining his intentions.

Also of obvious use are interview transcripts and videotapes in which artists comment on physical attributes, symbolic meaning, and viewer interaction.[17] These become especially useful when interviewers succeed in having artists express the core conceptual and aesthetic intention of their project, rather than just the surface details of proper execution. In this way statements by artists (and their associates) may serve as the basis for decisions, for instance about new technologies unimagined at the time of the interview. Photographs, video recordings of successive installations, and other visual aids can further clarify early artist statements about their work.[18]

V. Collaborators on hand

In addition to documentation, the conservator and other personnel 'on the ground' always affect the outcome of an installation. In regard to conventional art, each museum occupational specialty plays a more or less identifiable role, to a degree in sequence – e.g. registration, curation, conservation, and exhibit design. Depending on type of work and institutional culture, various specialists consult with one another and are physically co-present during various stages. But the process evolves around a routine, albeit with some variation based in genre-paintings, say, versus sculpture.

Installation art disrupts the typical sequencing, disturbing the usual boundaries and bringing in new actors – or at least old actors in new ways.[19] The curator carries the art historical eye – an understanding of the art in relation to art history, the work's social context, and its formal aesthetic properties. The registrar has knowledge of precisely what was acquired in the transaction, including contractual imperatives and constraints that dictate how it should be shown. The exhibition designer strives to fit the work into the physical context of the space and the structure of the exhibit. Preparators participate in getting the various elements together and on view in a safe and timely way.

Somewhat newer to the scene are technical specialists. New media installations need audio-visual and IT personnel both for initial display and for long-term survival. These people move beyond the role of technicians to become full-fledged professionals whose advice and concerns shape the work. Their expertise helps determine how much heat appliances will generate, the capacity to contain (or amplify) equipment noise, and the ability to recreate sound and visual effects through replacements of obsolete display equipment. Acoustic engineers may be called in to document and then simulate precise sound characteristics.

Most important among the new cast of characters are the artists and those who work directly with them, as a continuing force in the life of the artwork.[20] On scene during an installation, artists at least partially shift roles from original creator to collaborator as they re-negotiate the nature of the work in new circumstances and with other actors. They improvise artifact placement and equipment functioning – and decide in a collective act when the work is once again 'finished.'[21] Many installation artists expect, and even stipulate in their contracts, that they or their associates be involved with each re-installation of their work. Although collecting institutions appreciate artists' presence in re-installations, some eventually come to think they have acquired sufficient knowledge to re-install without the artist's presence (and daily rates). Otherwise, because of the dynamism of installation as art form, the owner is not so much buying a finished work as contracting an artist and their designees – in perpetuity. Such a development, perhaps appropriate and perhaps not, changes the nature of what it means to be a museum, what it means to be an artist, and what it means to be an artwork.

While the conservator has long played a role in shaping the ongoing life of artworks through their interventions, in the case of installation art that role is more ambiguous and potentially more intervening. The conservator may enter in debate about whether replacing unstable components destroys the meaning of the work, or whether the institution should even acquire an installation with a high cost of future equipment purchase and media migration.[22] Even if acquired, conservators may weigh in on whether to accession all of a work's components. Decisions not to acquire may allow institutions to commission installations for one-time display, giving the artist full or partial rights to re-create the piece in the future. Such negotiations, not widely practiced in conventional art production, can determine the identity and fate of an installation work.

An evolving role for conservation

Despite all the new-fangled aspects of installations, several old-fashioned conservation skills remain relevant. Most obviously, conservators know materials. And physical elements do make up much installation art, regardless of how esoteric the conceptual underpinnings. Even for artifacts meant to decay, conservators can facilitate having them decay the right way. The hodge-podge of hardware and media are, as Latour[23] would say, 'actants' that press the case for *doing something*. Conservators begin with the assumption that physical artifacts, like human beings, 'behave' in response to their environments. Conservators modulate environments to influence such behavior. Other fundamental conservation practices that carry over to these new works are systematic documentation, and research across art and science to arrive at considered options. Conservators are used to dealing with small details as well as the big picture.

While conservators' hands-on skills become less pertinent with the erosion of ideals of object-authenticity, the changed art scenario opens new ways to think of conservation's future. Rather than spelling the demise of conservation (why not let the artist, installers and AV department do it?), the contingencies at work invite new thinking about what it means to preserve, replicate and repeat. In suggesting the model of iteration, as Tina Fiske does in this volume, the relevant precedent shifts from artifact frozen in time towards precedents – as in musical and theatrical performance. There is documentation, like score or script that grounds the work within particular guidelines but does not presume identical renditions each time the work is re-created.[24] Successive renditions build, comment, and elaborate on what has come before. Variation becomes normal and expected, rather than a sign of imperfection. Without falling back to an unmitigated subjectivity, conservation can acknowledge the inevitability of contingency and accept, perhaps even embrace, complexity, change, and even some contradiction.[25]

Sustaining the work involves not a single type of expert but a collection of expert individuals, all of whom buy in to a particular yet evolving vision of what it consists. This likely means that each practitioner also changes by virtue of working in ensemble.

These new collaborations impact the internal structure of galleries and museums. The San Francisco Museum of Modern Art early on established a 'team media,' consisting of staff conservators, curators, registrars, exhibition technicians, and IT specialists to address the complex needs of their media installations. Responding to this model, Tate, MoMA and other museums have followed suit.

Still more ambitious modes of collaboration take shape across institutions and practitioners.[26] In contrast to conservation's craft background, which enabled a delimited set of skills – knowledge, say, of metal or paint – to be applied to a wide range of particular artifacts, the idiosyncrasies of installations make it harder for relevant expertise, even of just materiality, to cumulate in a single expert. Instead, useful knowledge becomes distributed across an array of practitioners. The trick becomes gathering it up and distributing it. To that end, the Netherlands Institute for Cultural Heritage and the European Union sponsor the International Network for the Conservation of Contemporary Art (INCCA).[27] INCCA programs link their network members – conservators, artists, art historians, scientists, registrars, and archivists – who share knowledge about artists and their work.

And this brings us to still another possible role for the conservator, that of coordinator. Without imposing a new rigidity among the art trades, it is at least worth suggesting that those whose backgrounds include material expertise, methods of documentation, art history, and a professional eye on the past and the future, can help bring together the diverse skills and orientations that quite literally make up the art. Whoever takes on the role must have – and here we are dealing with qualities that do not come bound up in the traditional conservation repertoire – sufficient understanding of interpersonal process to overcome the tensions that naturally arise among people who otherwise function in a more autonomous manner. The museum itself becomes less a collector of things and more a mechanism of collaboration and an arranger of experiences.

Some institutions now actively combat images of being elite guardians of cultural heritage and instead present themselves as the vibrant heart of the 'creative city.'[28] A growing number devote themselves to generating interest by dint of charm, spectacle, or surprise. In this context, installation projects have gained ground, raising the ante of conservation decisions that affect them. Quite ironically, given past anxiety about the threat to art posed by mechanical reproduction through photography or endless multiples, installation art comes along as something that insists on real-life experience of an artist-approved environment. Through sheer physicality as well as media heterogeneity, virtual representations cannot easily duplicate being there. The merry pranksters and artist activists who set out to make the museums irrelevant, or at least their cultural descendants, emerge as the museum's ace in the hole. No matter how easy and cheap to mount a digital Mona Lisa, only by actually going to the gallery can you find out what it means to crush the Guernica/Gernika or ingest a work by Felix Gonzalez-Torres. Somehow, and in ways only now beginning to take form, conservation's role will include the ability to sustain such revelations.

Acknowledgements

The authors thank Vivian van Saaze for her helpful comments.

Notes

1. Cornelia Weyer examines the difficulties of applying established restoration theory to installation art (Cornelia Weyer, "Restoration Theory Applied to Installation Art," *VDR-Beiträge zur Erhaltung von Kunst und Kulturgut* 2:2006, 40–48, 10 June 2007 <http://www.inside-installations.org/project/detail.php?r_id=136&ct=maastricht>).
2. Claire Bishop, *Installation Art: A Critical History* (London: Tate Publishing, 2005).
3. Riet De Leeuw, "The Precarious Reconstruction of Installations," *Modern Art: Who Cares?* eds., IJsbrand Hummelen and Dionne Sillé (Amsterdam: The Foundation for the Conservation of Modern Art and the Netherlands Institute for Cultural Heritage, 1999) 212–221.
4. Christian Scheidemann, "Gordon Matta-Clark's Object Legacy," *Gordon Matta-Clark: You Are the Measure*, ed. Elisabeth Sussman (New York: Whitney Museum of American Art, 2007) 119–123.
5. The "agency" of objects is now central to numbers of analytic efforts, most notably actor-network theory, see e.g. Bruno Latour, *Reassembling the Social: An Introduction to Actor-Network-Theory* (New York: Oxford University Press, 2005).
6. A complex project, of any sort, says social theorist Bruno Latour, survives through multiple actors' positive vision of its completion – their "love" for its reality. See Bruno Latour, *Aramis or the Love of Technology* (Cambridge: Harvard University Press, 1996).

7. William A. Real, "Toward Guidelines for Practice in the Preservation and Documentation of Technology-Based Installation Art," *Journal of the American Institute for Conservation*, Volume 40 (2001): 218.
8. Emulating computer art was the topic of a Guggenheim Museum symposium *Echoes of Art: Emulation As a Preservation Strategy* 8 May 2004. Transcripts are posted: <http://www.variablemedia.net/e/echoes/index.html> (Accessed 1 November 2007).
9. Martha Buskirk, *The Contingent Object of Contemporary Art* (Cambridge, MA: Massachusetts, Institute of Technology Press, 2003).
10. Chrissie Iles and Henreitte Huldisch, "Keeping Time: On Collecting Film and Video Art in the Museum," *Collecting the New: Museums and Contemporary Art*, ed., Bruce Altshuler (Princeton and Oxford: Princeton University Press, 2005) 65–83.
11. Justin Graham and Jill Sterrett, "An Institutional Approach to the Collections Care of Electronic Art," *WAAC Newsletter* 3 September 1997: 19, accessed 10 June 2007, <http://palimpsest.stanford.edu/waac/wn/wn19/wn19-3/wn19-310.html>.
12. Kira Perov, "Interview with Kira Perov". Electronic Arts Intermix Online Resource Guide for Exhibiting, Collecting & Preserving Media Art, 17 May 2006 (*http://resourceguide.eai.org/preservation/installation/interview_perov.html*) (Accessed 10 June 2007); Bill Viola, "Permanent Impermanence," Mortality/Immortality? The Legacy of 20th-Century Art, ed., M.A. Corzo (Los Angeles: Getty Conservation Institute, 1999) 85–94.
13. Sanneke Stigter, "Certificate of Glass (one and three)" 2007, Inside Installations, 10 June 2007 (http://www.inside-installations.org/artworks/detail.php?r_id=441&ct=research).
14. John Hanhardt, Excerpts from Conference "Preserving the Immaterial: A Conference on Variable Media," which took place at the Solomon R. Guggenheim Museum, New York, on 30–31 March 2001, *Permanence Through Change: The Variable Media Approach* (New York: Guggenheim Museum Publications, 2003) 74–77.
15. John Bowser, Art Programs Administrator, Dia Foundation, personal interview, 14 May 2007.
16. Variable Media Network (http://www.variablemedia.net/) (Accessed 1 November 2007).
17. Carol Mancusi-Ungaro, "Original Intent: The Artist's Voice," in IJsbrand Hummelen and Dionne Sillé, eds., *Modern Art: Who Cares?* (Amsterdam: The Foundation for the Conservation of Modern Art and the Netherlands Institute for Cultural Heritage, 1999) 392–393. The International Network for the Conservation of Contemporary Art publishes guides for conservation interviews with artists in their methodology section (http://www.incca.org/) (Accessed 1 November 2007).
18. Carol Stringari, "Installations and Problems of Preservation", *Modern Art: Who Cares?*, IJsbrand Hummelen and Dionne Sillé, eds., (Amsterdam: The Foundation for the Conservation of Modern Art and the Netherlands Institute for Cultural Heritage, 1999) 272–281. Also, Real recommends multiple documentation formats that incorporate different points of view. He references Gary Hill's concern that some documentation, such as a photograph, may sanctify and accentuate details that are actually irrelevant to the piece (William A. Real, "Toward Guidelines for Practice in the Preservation and Documentation of Technology-Based Installation Art," *Journal of the American Institute for Conservation*, Volume 40 (2001): 221). In addition, Laurenson provides a model for documenting a complex video installation

by Bruce Nauman (Pip Laurenson, "Inside Installations: Mapping the Studio II," 2006, *Tate Research*, accessed 10 June 2007 (http://www.tate.org.uk/research/tateresearch/majorprojects/nauman/themes_3.htm).

19. IJsbrand Hummelen, *Conservation Strategies for Modern & Contemporary Art*, 2005, Inside Installations, 10 June 2007 (http://www.inside-installations.org/OCMT/mydocs/HUMMELEN%20Conservation%20Strategies%20for%20Modern%20and%20Contemporary%20Art.pdf).

20. Laurenson says about time-based media, but is relevant to a still wider class of installations, 'Authorship is maintained by the causal link to the artist and the properties that the artist considers mandatory.' (Pip Laurenson, "Authenticity, Change and Loss in the Conservation of Time-Based Media Installations," *Tate Papers*, Autumn (2006), accessed 10 June 2007 (http://www.tate.org.uk/research/tateresearch/tatepapers/06autumn/laurenson.htm).

21. The completion of any artwork always has a greater ambiguity than is commonly recognized (Howard S. Becker, Robert R. Faulkner and Barbara Kirshenblatt-Gimblett, Art from Start to Finish [Chicago: The University of Chicago Press, 2006]), but with installation art, the finish is more radically indeterminate – and collective.

22. In a parallel discussion, Laurenson offers important insights into the conservator's role in time-based media (Pip Laurenson, "Authenticity, Change and Loss in the Conservation of Time-Based Media Installations," *Tate Papers*, Autumn (2006), accessed 10 June 2007 (http://www.tate.org.uk/research/tateresearch/tatepapers/06autumn/laurenson.htm).

23. Bruno Latour, *Reassembling the Social: An Introduction to Actor-Network Theory* (New York: Oxford University Press, 2005).

24. Pip Laurenson, "Authenticity, Change and Loss in the Conservation of Time-Based Media Installations," *Tate Papers*, Autumn (2006), accessed 10 June 2007 (http://www.tate.org.uk/research/tateresearch/tatepapers/06autumn/laurenson.htm); Bill Viola, "Permanent Impermanence," *Mortality/Immortality? The Legacy of 20th-Century Art*, ed., M.A. Corzo (Los Angeles: Getty Conservation Institute, 1999) 85–94.

25. For instance, The European Union funded thirty-three case studies and theoretical papers to research preserving installation art in European collections (http://www.inside-installations.org) (Accessed 1 November 2007). In another instance, the Kramlich Foundation, Museum of Modern Art (New York), San Francisco Museum of Modern Art, and the Tate created Media Matters that generates best practice documents for managing time-based media installations (http://www.tate.org.uk/research/tateresearch/majorprojects/mediamatters/) (Accessed 1 November 2007).

26. For the idea that art, or at least architecture including "great" architecture, contains contradictory and contingent elements, see Robert Venturi, *Complexity and Contradiction in Architecture* (New York: Museum of Modern Art, 1977).

27. The International Network for the Conservation of Contemporary Art (http://www.incca.org) (Accessed 1 November 2007).

28. See Jules Lubbock, "Tate Modern in the Age of e-Production," *Visual Culture in Britain Research Group*, Volume 4:2, (2003), 2007; Richard Florida, *Cities and the Creative Class* (New York: Routledge, 2005); Sharon Zukin, *The Cultures of Cities* (Oxford: Blackwell, 1996).

20

Contemporary Museums of Contemporary Art

Jill Sterrett

It is a familiar scenario in a museum of contemporary art. It is mere days before the opening. An artist is working on-site on a new installation and, as always, it has a way of stirring things up. Everyone wonders if it will come together for the opening, knowing somehow it always does. There is genuine satisfaction, albeit ragged, among all involved – art handlers, technicians, curators, registrars and conservators. What results is spectacular and the rumblings heard next are all too familiar. *Let's keep it*. These three words set waves of new activity in motion. The energies that support an artist's initial installation are not the same as those that will support keeping it.

Contemporary art is about now and, as such, museums of contemporary art are called upon to keep pace. To operate in the present means to value agility; keeping current is, after all, key to being about whom we are today. Default to this brisk tempo, however, and there is the real risk of paving over a host of observations and details that will be critical to ongoing and future interpretations of the art of our time. It is within this aspect of the museum's purview that a slower process of looking and looking again is rewarded. What museums of contemporary art really seek is a mechanism for variable speed.[1] And, at the heart of variable speed operability lie two questions – what takes time and what takes the passage of time?

To answer the first question is to tackle the tensions that arise from too few minutes in a day and in considering this it is critical to acknowledge that the larger ecology of museums – an ecology which embraces the public, artists, educational institutions and galleries – matters.[2] In 2005, the Rand Corporation reported on the shifting patterns in Americans' leisure time and tastes noting that museums are in competition with other recreation and entertainment industries. Diverse populations and new patterns of funding have led to observable shifts in the way arts organizations operate. Targeting and attracting audiences, managing resources, and securing funding are all described as priorities.[3] This competition with movies and concerts is manifest in many ways within museums but is certainly evident in the breakneck pace with which art and exhibitions go

Conservation: Principles, Dilemmas and Uncomfortable Truths

up and come down. Dark galleries are frowned upon and installation times are compressed.

Further, the question of what takes time begs a closer look at contemporary art forms and the mechanics of installing and caring for them. Museums grasp the time needed to conserve a painting but not necessarily the commitment of time needed to keep a multi-part, large-scale installation. Why is that? Perhaps it is the material diversity that contributes to genuine confusion within the museum about who is supposed to do what. Maybe the actions of caring for complex works of contemporary art coincide with their installation and deinstallation in the galleries – as opposed to the sequestered confines of a conservation studio – and this shifts the rhythms of traditional museum practice. Take for example, the film, video, photography, self-lubricating plastic, Vaseline and salt that Matthew Barney customarily employs. Or, consider sculptural works like *Things Fall Apart* by Sarah Sze made of the parts of a Jeep Cherokee, packing peanuts, bottle caps, balsa wood, mai tai umbrellas, and aspirin to name just a few of her materials (Figure 20.1). We rely on Barney and Sze, and their assistants, to realize the works and to orchestrate what goes on in the galleries, deferring the museum's command for a later date. Making the parts become the whole, learning how to

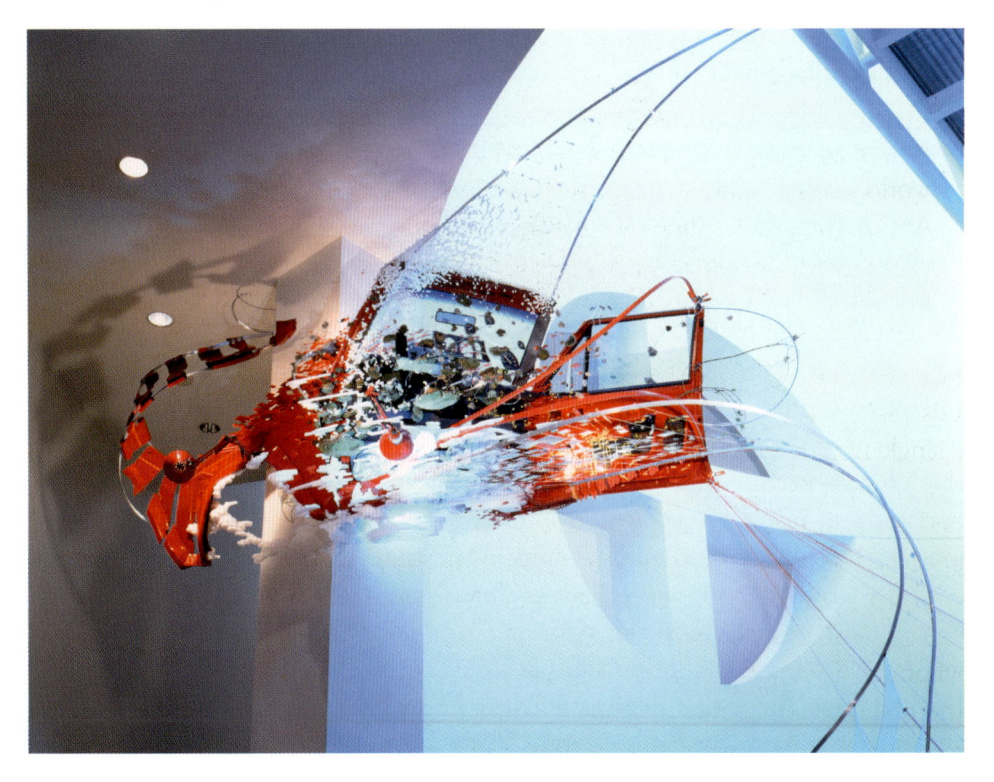

Figure 20.1 *Sarah Sze,* Things Fall Apart, *2001, mixed media installation with vehicle. Collection SFMOMA, Accessions Committee Fund. Copyright Sarah Sze. Photo reproduced with permission from Frank Oudeman.*

do this from the artists, creating the documentary tools for the future – the bottom line is caring for contemporary art takes more time than we give it during installation and deinstallation.

What about the second question? What about the passage of time? This is not minutes and seconds on the clock. This is time with the long view in mind. In a museum, it is the notion of temporality that situates art within a context; approaches it and describes it as part of a larger, discursive continuum. Indeed, this is a perspective of time upon which one of the museum's most revered attributes – that is trust – rests. In 2001, the American Association of Museums conducted a survey on trust; in particular, the trust the American public has in museums as objective sources of information. Eighty-seven per cent of people surveyed reported having trust in museums. Thirty-eight per cent trusted museums the most.[4] Howard Fox, curator at the Los Angeles County Museum of Art, wrote about the curatorial conflict embedded in these results. In his words, 'For however buoying the public's confidence may be, it may ultimately serve to confine and frustrate the real work that curators – and particularly curators of contemporary art – do.'[5]

> In today's global art world, which flourishes in societies that tolerate divergent cultural traditions and ideologies, museums have a responsibility not to serve as ideologues for the hegemony of one cultural vision over all others. In encyclopedic museums especially, a cosmopolitan spirit of intercultural exploration is appropriate. And an awareness of how quickly the art world evolves, always in relation to the larger world beyond, is essential. Anticipation of the future, rather than codification of the past, is a necessary attribute of the contemporary curator's function. This aspiration is anything but certain, its expression is anything but authoritative.[6]

Fox makes a pitch for fallibility and the case for the curatorial right to be wrong. His case signals two shifts for curators in museums: first, he rejects the notion of a single narrative – the idea of a definitive collection that imparts objective truth – describing it as 'wrong-headed in conception and distracting in practice' and, second, he calls out the very real disconnect between exploring works of today and, at the same moment, assigning importance to them for all time.[7]

This perspective resonates beyond the curatorial realm but it manifests in a slightly different form for a range of other museum colleagues; colleagues for whom it is not so much about the right to be wrong, as it is about the right, and the obligation, to rethink. Keeping contemporary art is a challenge that calls for retooling our standard methods of care. For example, the traditions of conservation still hold that preserving a work's integrity is the bedrock of practice and that a work's integrity is linked to its true nature at the moment of creation. With this in mind, the last fifty years of art making is bound to have a destabilizing effect.

Whether it is the unorthodox and transient materials such as soot, lard and lettuce found in Arte Povera, the canvases in Robert Rauschenberg's *White Painting* series which have been painted and repainted white by an assortment of different hands since they were originally made in 1951, or Felix Gonzales-Torres' ever generous inclusion of the public in his candy spills and stack pieces, installations and other contemporary works often confound what constitutes the finished work of art. By this measure, linking the integrity of a work of art to its nature at the moment of creation hems conservators and collection stewards in by their own version of distracting wrong-headedness.

The good news is that active discourse in recent years is transforming conservation methods for contemporary art.[8] Increasing sophistication in theoretical and practical problem-solving leads us now to examine whether museums, as the tools of understanding that we aspire them to be, are actually equipped to operate in the ways the art needs them to. In *The Contingent Object of Contemporary Art*, Martha Buskirk observes that 'the transition from a work of art's initial appearance to its extended life as an object to be preserved, collected, and contextualized as part of a historical narrative involves a complex process of negotiation.'[9] Are museums attuned intellectually, spiritually, practically, and financially for this negotiation?

Museums of contemporary art must reckon with the projective possibilities of the art's unwritten history and, at the same time, continue to carry the mantle of the public's trust. Fox's assertion that curating contemporary art would benefit from a more exploratory tone, rather than an authoritative one, points the entire museum in a promising direction. Consider the shift for conservation. This is an approach that acknowledges that, with contemporary art, there are aspects that are certain and others that are not so clear. It is an approach that recognizes that there are things we know quickly and things we cannot; an approach that values learning and relearning, seeing and seeing again over time. What is fixed and what is variable? It is a question derived from the Guggenheim's initiative on the legacy of contemporary art and it has proven to be vital as an organizing principle for stewardship of the art of our times.[10] Between the archival, factual accounting of a particular work and its realization in a given exhibition, there's an interpretive gap.[11] In fact, the complex negotiation between a work's initial appearance and its extended life distills down to mediating its variability in this gap.

The challenge of managing variability in contemporary art is a driving force behind shifts in the way museums maneuver. For one, contemporary art calls for the involvement of an entire corps of museum experts. As the works themselves commingle the physical with the conceptual, the virtual and the contextual, there's a corollary commingling of custodial roles. More than that, wholly new expertise is required, from IP lawyers and webmasters to artists and their assistants. Contemporary art museums rely more and more on the efforts of expert teams. Secondly, the artist's role has changed. With contemporary works comprised of multiple, and often variable, components, understanding builds with each instance of display and this understanding profits from input from the artist. In this way, artists do more than create their work; museums rely on their ongoing involvement as part of a strategic hand-off of the work. Artists are engaged with their work over time and, in so doing, their charge as creator extends and expands into the realm of steward. Finally, this hand-off of the work over time through a series of occurrences makes each instance of decision-making distinctly open-ended, perhaps uncomfortably open-ended given the traditions of museums. Science has contributed to the development of highly refined preventive methods of care for traditional forms of art that ensure longevity on the order of hundreds of years, as long as display is restricted. With contemporary art forms, a particular preservation solution may only have a 5–10 year horizon and reinstallation is the vehicle for outlining a plan for the next horizon. Take, for example, the scores of works for which replacement of ephemeral materials or rapidly obsolescent technologies is required and the inevitable conclusion is the inherent need for frequent display. Keeping contemporary art often means displaying it.

Archeologists have a term called a *find*. To discover a find is to discover something that may convey aesthetic value but it may be just as notable for its information value.[12] Applied in reverse, it is a particularly appealing description of what we might aspire to do in museums of contemporary art; namely, to plant finds. These are the clues that trace our ongoing engagement with a work, clues that reveal its life over time. Planting finds may be the answer to addressing the range of variables in contemporary art and, in this way, the cornerstone of Buskirk's complex process of negotiation. The concept is appealing because it adjusts the burdensome tone of authority museums inherit as sources of objective truth by actively committing to seeing and seeing anew over time. To pull this off, museums of contemporary art are called upon to cultivate, among other things, ways of maneuvering with variable speed; rapid cycles of engagement paired with sustained follow-through, in-the-moment presence coupled with reflection, breakneck speeds that work in tandem with strategic pauses.

Taking into account the transitory nature of ephemeral materials, built-in physical variability and the performative elements that characterize so much of the art of the last fifty years, the work of a contemporary art museum is not business as usual. When it works well, the result is a finely tuned machine that reinforces our collective commitment to art and to history and when it works really well the rewards may be even more far-reaching because we contribute to the ways in which museums everywhere are transforming in the twenty-first century.

Notes

1. Dominic Willsdon, Leanne and George Roberts Curator of Education and Public Programs, San Francisco Museum of Modern Art, conversation with the author, November 2006.

2. Saskia Sassen, "The City: Strategic Space/New Frontier," ISEA 2006 Symposium, San Jose California, August 2006.
3. Kevin F. McCarthy, Elizabeth H. Ondaatje, Arthur Brooks, András Szántó, *A Portrait of the Visual Arts Meeting, Meeting the Challenges of a New Era* (Santa Monica: Rand Corporation, 2005) xviii, 83–85.
4. American Association of Museum, "AAM Press Release: Americans Identify a Source of Information They can Really Trust," 7 May 2001 (http://www.aam-us.org/pressreleases.cfm?mode=list&id=21) (accessed 15 August 2007).
5. Bruce Altshuler, ed., *Collecting the New: Museums and Contemporary Art* (Princeton University Press, 2005) 17.
6. Altshuler, p. 26.
7. Altshuler, p. 25.
8. Notable published efforts include: Miguel Angel Corzo, ed., *Mortality/Immortality?: The Legacy of 20th Century Art* (Los Angeles: Getty Conservation Institute, 1999); IJsbrand Hummelen, and Donne Sille, eds., *Modern Art: Who Cares?* (Amsterdam: Netherlands Institute for Cultural Heritage, 1999); *Media Matters: Collaborating Towards the Care of Time-based Media* (a consortium project of the New Art Trust, Museum of Modern Art, San Francisco Museum of Modern Art and Tate), (www.tate.org.uk/research/tateresearch/majorprojects/mediamatters).
9. Martha Buskirk, *The Contingent Object of Contemporary Art* (Cambridge: Massachusetts Institute of Technology Press, 2003) 12.
10. Alain Depocas, Jon Ippolito, and Kaitlin Jones, eds., *Permanence Through Change: The Variable Media Approach* (Montreal: Daniel Langlois Foundation, 2003) (http://variablemedia.net).
11. Nancy Troy, Professor of Art History, University of Southern California, in discussion at "Object in Transition: A Cross-Disciplinary Conference on the Preservation and Study of Modern and Contemporary Art," The Getty, January 25, 2008.
12. Nicholas Stanley Price, M. Kirby Talley Jr., Allessandra Melucco Vaccaro, eds., *Historical and Philosophical Issues in the Conservation of Cultural Heritage* (Los Angeles: Getty Conservation Institute, 1996) 210.

21

White Walls: Installations, Absence, Iteration and Difference

Tina Fiske

In May 2007, Andy Goldsworthy and a team of assistants installed *White Walls* over a period of five days.[1] A large room-sized installation, it comprised 18,200 lbs (8255 kg) of wet porcelain clay, rolled in one inch thick slabs and applied to the walls to form a smooth continuous coating that covered the total wall surface area of Galerie Lelong's main gallery (some 1975 square feet) (Figure 21.1). Like other Goldsworthy's clay installations, *White Walls* resulted from a very particularized and intensive installation process involving many participants, of which the work itself becomes a material embodiment. However, unlike other clay rooms that he has built, this one proceeded to de-install itself – an outcome intended by the artist – but one that occurred more decisively than he had fully expected. Applied wet and directly to the gallery walls, with no additional substructure or grip, Goldsworthy intended for the clay to dry and crack, and then gradually delaminate onto the floor. In advance of the installation, Goldsworthy had felt unable to predict how the drying, cracking and delaminating would actually proceed, but he had expected that the clay would dry and crack on the walls first, before falling.

As it happened, the clay fell during the drying process; in fact, beginning almost immediately during the first night. Arriving at Galerie Lelong the next day, a large swag of damp clay had dislodged from a short, newly installed (and hollow) partition wall. The clay had not only stripped away two freshly applied coats of white paint, but also the surface of the Sheetrock (from which the wall was constructed).[2] Over the next five days, the contraction of the clay and formation of cracks dislodged 95 per cent of the clay covering, which fell quite dramatically and noisily in large damp sections that ultimately dried on the floor (Figure 21.2). For those in the room, midst the falling clay, the dull thud was quite shocking, but ultimately compelling. The various sections and slabs wrought not inconsiderable destruction, taking with them five and a half years' worth of paint layers, revealing the physical substance of the walls and the traces of previous interventions by other artists (Figure 21.3).

Goldsworthy initially conceived *White Walls* in such a way that the clay could be fairly easily re-processed and recycled, and the installation re-staged by

Figure 21.1 *Andy Goldsworthy,* White Walls *(Day 1), 2007. Reproduced with permission from Galerie Lelong, New York.*

Figure 21.2 *Andy Goldsworthy,* White Walls *(Day 3), 2007. Reproduced with permission from Galerie Lelong, New York.*

White Walls: Installations, Absence, Iteration and Difference

Figure 21.3 *Andy Goldsworthy,* White Walls (Day 16), *2007. Reproduced with permission from Galerie Lelong, New York.*

others in the future to his specifications.[3] Re-using the same clay for future incarnations, or at least those of the immediate future, would ensure a material kinship between realizations, and allow the memory of activity and human presence to accumulate. However, the recalcitrance that seemingly manifested in the work's performance at Galerie Lelong in the form of its radical and rapid delamination prompted Goldsworthy to modify that preliminary opinion. A large proportion of the clay fell from the wall with sections of both white and black paint attached – rendering it unusable for the purposes of reinstalling the work. For me, what proved fascinating about *White Walls* was the installation's volition, the way it appeared to break with Goldsworthy's specifications and, moreover, with itself, at least in the respect that we might take it to constitute a document of its own making. In view of the environmental contingencies of the Lelong space, installing *White Walls* elsewhere, subject to other environmental conditions, could issue in a very different *White Walls*. What if, in a subsequent context, for instance, all of the clay remained adhered to its host walls? If it did, how would it relate to the first realization, wherein the dramatic manner of its de-installation became so definitively part of the work? Installed in another context, *White Walls* could exhibit profound differences, but it seems to me that currently there is little critical way of accounting for differences that arise between incarnations of a work, or of the role played in that respect by absence or rupture. A default position might be to ask whether a future, non-delaminating *White Walls* would 'become another work.' I must confess I increasingly find this to be too anodyne a response to the issue of difference. What *White Walls*' gradual process of desiccation and its apparent dissociative behaviour (both from the literal surface of the wall, from Goldsworthy's expectations, and also from itself) brought to my mind is philosopher and linguist Jacques Derrida's notion of 'iterability,'[4] which is a particular mode of repetition that mobilizes notions of breach, absence and difference.

Derrida referred to iterability as 'the possibility of . . . being repeated (another) time.' It is, however, a possibility that 'alters, divides, expropriates, contaminates what it identifies and enables to repeat itself.'[5] Of particular interest to me are the structural roles that it accords to absence, rupture, and to difference within the process of repetition. These concepts are rarely explicitly deployed in conservation discourse. Various commentators, such as Jill Sterrett and Martha Buskirk, acknowledge 'incompleteness' and 'intermittance' as characteristics of installation, and as critical factors bearing upon the task of re-installation.[6] Yet there remains a preference for a vocabulary of 'presence,' 'identity' and 'actualizing.' In this short essay, I want to invoke iterability and deconstructive notions such as absence, rupture and difference as a way that challenges discussions of re-installation.[7] What radical propositions might they hold for thinking about installations such as *White Walls*, and how their first de-installed and re-installed incarnations stand in relation to each other? The putative 'paradigm shift' that much contemporary art has necessitated of the conservation discipline notwithstanding,[8] I contend that conservation has not ultimately unpacked either its retention of 'presence' as a presiding value, or the further inference that it, as a discipline, constitutes a mode of domestication. I point to literary translation, which has, in the last thirty years, undergone a significant discipline challenge, and might provide models that enable conservation to recast its critical purchase vis-à-vis installations and the task of their recreation. More broadly, it seems increasingly crucial to address the experience of alterity[9] that installations conceivably prompt in conservators, and whether conservation might broach something akin to an ethics of alterity or otherness.

* * *

> In order for the tethering to the source to occur, what must be retained is the absolute singularity of a signature-event and a signature-form: the pure reproducibility of a pure event (Derrida, 1988, p. 20).

Whilst many conservators of installation works of art would agree that 'the pure reproducibility of a pure event' constitutes an improbable or misguided goal, the issue of 'tethering' any one installation to its respective 'source' has featured significantly in recent discussion on the long-term viability of installation artworks, and has converged particularly around the issue of 'variability.'[10] Critical in this endeavour is the emphasis placed upon the 'authorized' version or incarnation of a particular work, and upon the person of the artist. Authority, where it is designated upon a specific version of an installation, operates to ensure of the work 'the absolute singularity of a signature-event and a signature-form.' Meanwhile, documents, interviews and specifications very often act as written supplements that secure the 'non-presence' of the artist. In general, the presumption that installation is a particular mode of artwork (and we might consider *White Walls* as an eminent candidate) in need of tethering is increasingly left unchallenged. The need for tethering arises, of course, because of the lack of a persisting material object, which would typically maintain a physical bond to its originating event or context of inscription, and to which it would offer us privileged access. With installations, tethering becomes a prerequisite precisely because of the imminence of de-installation, and the possible ambiguities that an intermediate or disassembled state can produce for a work. Tethering secures the work-in-absentia, disarming absence as a condition that could threaten the viability of the work, and rendering it essentially benign.

What, however, about *White Walls*? To what extent could its manner of de-installation propose a more radical understanding of its current state of de-installation? In his 1971 paper, *Signature Event Context*, Derrida elaborated on writing as a mode of communication underwritten by absence; the written word often emerges, he argued, in contrast to the spoken word, in the physical absence of the addressee, and is typically read in the absence of the writer. Derrida recognized writing as a communicative mode that becomes radically unbound from its 'real' context or moment of production by virtue of a breaking force or 'force de rupture' (Derrida, 1988, p. 9). This rupture is not, he contended, an accidental effect, but part of the structure of writing, and predicated on 'a certain element of play, a certain remove, a certain degree of independence with regard to the origin, to production, or to intention in all its "vital", "simple", "actuality" or "determinateness"' (Derrida, 1988, p. 64). Not least because of the way that it can be seen to have literalized this rupturing momentum, *White Walls* has prompted me to consider that installations could have that same breaking force as a structural capacity; this is not least because they are underwritten by their own imminent material dis-assemblage, that is by the fact of their pending de-installation and subsequent absence. As Derrida himself emphasized, 'if a certain "break" is always possible, that with which it breaks must necessarily bear the mark of this possibility inscribed in its structure' (Derrida, 1988, p. 64). Where, however, does this leave the process of tethering, and within that, the prominent role often afforded the artist in determining the future shape of their work?

Some areas of recent discourse have attempted to mitigate what we might take to be 'force de rupture' in installation. Much has been made of an analogy to musical scoring and re-performance, and to philosopher Nelson Goodman's distinction between 'autographic' and 'allographic' works of art specifically, particularly insofar as the latter classifications involve 'the use of notational systems that allow the proliferation of multiple legitimate instances of a work.'[11] Appeals by conservators such as Pip Laurenson to Goodman's allographic classification draw upon its two-part basis, and the unequivocal bond that Goodman's distinction sets up between those two parts. The first of those is set down by the author at a given time and place, using a recognized or conventionalized system or template that allows the work to be actualized beyond those originating givens – by a second, third, fourth, fifth participant – a 'signature form' that enables the instantiation of the 'signature event.' Laurenson elaborates an allographic reading of 'time-based media installations,' which she argues, 'exist on the ontological continuum somewhere between performance and sculpture. They are similar to works that are performed, in that they belong to the class of works of art, which are created in a two-stage process' (2006). Where she refers to Stephen Davies' *Musical Works and Performances: A Philosophical Exploration* (2004), she does acknowledges that Davies takes the musical score to be 'ontologically significant.' Although Laurenson resists pushing the ontological significance of artists' specifications, there is a soft, perhaps unintended inference to that effect, and she employs adjectives such as 'thickly' or 'thinly,' derived from Davies, to describe the extent to which any one artist inscribes their instructions and preferences for future reinstallations of their work. That is notwithstanding various caveats: 'the installation is richer than the specifications . . . works that are performed allow for a greater degree of variation in the form that they take' (2006). Consequently, the conservator's ethical remit becomes focused on minimizing the erosion of identity between instances of a work.

My main contention with Laurenson's argument is that it is underwritten by an evident binarism, which she clearly draws from Goodman. She claims a 'conceptual dependency between the ontological framework in which an object is classified and described and the attending concept of authenticity.' Should the 'ontological framework' shift, then so too should our concepts of authenticity (2006). For Derrida, Deconstruction would seek to propose differing conditions than the kind of explicit binaries that Laurenson employs: 'Deconstruction does not consist in moving from one concept to another, but in reversing and displacing a conceptual order as well as the non-conceptual order with which it is articulated' (Derrida, 1988, p. 21). An enduring critical focus for Derrida was what he called 'the telos of "fulfilment",' which he argued underwrites disciplines that put intention at their centre. Derrida believed that fulfilment or plenitude are the end point (or 'telos') that all intentional expression tends towards, and with which, by extension, practices that are involved in actualizing intention get bound up (Derrida, 1988, p. 128). That latter impulse is explicit, I believe, with allographic readings of installation and the possibilities it offers for 'the proliferation of multiple legitimate instances of a work.' The 'legitimate instance' presupposes a notion of fulfilment, or of actualization. Derrida's iterability is a mode of repetition that, rather than 'aspiring to the fulfilment of the original,' searches or reaches beyond the original itself.

* * *

> The structure of iteration – and this is another of its decisive traits – implies *both* identity *and* difference (Derrida, 1988, p. 53).

So, as it stands, at the time of writing this essay the process by which *White Walls* could be recreated by individuals other than the artist is not documented; at least not as a set of instructions to which Goldsworthy has put his signature. The work is, to all intents and purposes, absent, and my questions remain: Re-installed in another context, is *White Walls* to be realized again as 'another work?'. Or, alternatively, how is the 'break' in *White Walls* to be re-enabled, issuing possibly in a work that does not break or delaminate? Would such a *White Walls* be viewed as weak or derivative in relation to its precursor? How, in effect, do we deal with difference as it might obtain in and between incarnations or 'iterations' of a work? Indeed, I wonder if we might think of the set of practices (de-installation, documentation, re-installation and so on) that keep any one installation viable as 'the movement of *différance*,' in contrast to the movement bound up with the pursuit of fulfilment or actualization.[12] Derrida described *différance* as 'the disappearance of any originary presence,' a disappearance that plays out with *White Walls* and inscribes the possible terms of its re-appearance: 'What is, is not what it is, identical and identical to itself, unique, unless it *adds to itself* the possibility of being *repeated* . . . and its identity is hollowed out by that addition, withdraws itself in the supplement that presents it.'[13]

Thomas Baldwin has recently reminded us that Derrida was engaged in what he referred to as a 'deconstruction of presence.' Concentrating on this particular phrase of Derrida's, Baldwin has suggested that Derrida did not want to deny the importance of presence 'conceived simply as our experience of things,' but rather what he (Derrida) referred to as 'self-presence.'[14] Baldwin paraphrases this as 'presence as an immediate and self-sufficient source of meaning . . . which is distinctly privileged in comparison to absence,' and suggests further that:

> it will become clear that our experience of the world needs to be understood in the context of a much more complex network of relationships involving things which are absent. So if anything is fundamental here, it is not presence by itself but the 'play' of presence and absence (2008, p. 111).

Exploring the 'internal relationship' between presence and absence has not rigorously emerged within discussions of re-installation. Yet I argue (occasioned by *White Walls*) that such discussions must engage with/in a 'play of presence and absence,' not least where installations pass a significant part of their 'existence' as 'things which are absent.' It is this that would permit us to think seriously of a notion of the 'minimal remainder' or a 'minimum of idealization' in relation to the re-installed or iterated work, both of which Derrida claimed iterability presupposes. As he contends, 'iterability supposes a minimal remainder . . . in order that the identity of the selfsame be repeatable and identifiable, in, through, and even in view of its alteration' (1988, p. 53). Under such conditions, a re-installed work would, in Derrida's words, never be that of 'a full or fulfilling presence.' As a remainder, it would be 'a differential structure escaping the logic of presence or the (simple or dialectical) opposition of presence and absence, upon which opposition the idea of permanence depends' (1988, p. 53).

Yet, if the re-installed work (a future *White Walls*, for instance) might thus be able to imply 'both identity and difference,' and avow itself as a minimal remainder or as an iteration that escapes the 'logic of presence,' where does this place the conservator's ethical injunction? Curator Jennifer Mundy once asked, 'Is it adequate, and intellectually coherent, to adopt one set of concerns and scruples for art works that adhere broadly to one view of the art object, and another set of different concerns and scruples for art works in which the object is dematerialized?'[15] A point of contention of Mundy's is the burdens produced by 'two different systems of decision-making,' and the manner in which they seem to have come to 'simply, normally coexist.' Yet, where, indeed, the call for a 'paradigm shift' has in some senses resolved into the 'simple' and 'normal' co-existence of two systems, much of the recent literature on the preservation of installations does not itself escape the 'logic of presence.' Whilst I argue that *White Walls* and installations more generally should be understood to escape or even to negate that logic, is it possible for the conservator to do so?

To that effect, I would like to see the dilemma of the conservator, when tasked with the re-realization of an installation, juxtaposed with that experienced by the translator when faced with a foreign-language text. Literary translation offers an appealing model for conservation on a number of levels, not least in what it conceives its 'source' or 'foreign' text to be or the status accorded to the translated text, but also in commonalities that exist between the history of translation and conservation practices, and shared values such as 'fluency,' 'transparency,' and deferred subjectivity. I have discussed sympathies and analogies between literary translation and conservation elsewhere.[16] Perhaps most pertinently, however, literary translation has in recent decades undergone something of a discipline challenge, particularly in respect of what translator and theorist Philip E. Lewis has termed the 'contradictory exigency' that constitutes 'the classical translator's predicament.'[17] Where the act of translation and its ethical injunction has been traditionally formed around the desire to be 'faithful both to the language/message of the original and to the message-orienting cast of [the translating] language,' theorist Lawrence Venuti indicates that translation has been revealed to be less a search for equivalence between these positions than it is a means to 'bring back a cultural other as the same, the recognizable, even the familiar.'[18] Translation is, Venuti suggests, 'the forcible replacement of the linguistic and cultural difference of the foreign text with a text that will be intelligible to the target-language reader' (2006, p. 18). This ineluctable recognition has sat at the core of the most dynamic thinking about translation ethics for several years, not least in how the translator recognizes, values or approaches difference.

How might looking at translation ethics direct the conservator to recognize and avoid effacing difference, not least where it is generated by the periods of absence that arguably underwrite installations and certainly punctuate their iterations? Moreover, how might translation ethics illuminate the experience of alterity or otherness that installations would seem to induce in the conservator? Derrida himself deployed the term 'anethical' to locate conditions such as alterity at the margins of what he called a 'given ethics' (1988, p. 122). Rather than view marginal conditions as 'anti-ethical,' we should take them to be anethical, and 'no less essential to ethics in general . . .' (1988, p. 122). More recently, literary theorist Shane Weller has noted that Derrida, amongst others, underestimated the difficulties in 'determining whether alterity is there at all, and if so, where.'[19] Weller has elaborated what he refers to as a 'space of the anethical,' which constitutes a 'kind of indecision' that might function to 'unsettle, if only in passing, any sense of where value might lie, and of what is ethical and unethical in any act of translation, however respectful or disrespectful that act might appear to be' (2006, p. 56). It is, he suggests, 'to be understood . . . as neither an ethics nor an alternative to ethics, but rather as a failure either to establish or negate the difference between the ethical and the unethical . . .' (2006, pp. 194–195).

Philip Lewis himself elaborated a translation strategy that would simultaneously 'force' the 'conceptual system of which it is a dependent,' and direct 'a critical thrust back towards the text that it translates and in relation to which it becomes a kind of unsettling aftermath.'[20] As such, his 'abusive fidelity' proposes an approach that enables the individual translator to take a position that speaks to his or her discipline as well as to the work in question. This strikes me as the kind of dual demand that installations in general and *White Walls* in particular precipitate, and to which the conservation discipline needs to direct itself. To view *White Walls* and other installations as iterable, as underwritten by absence, and to configure re-installation as a mode of iteration, offers this potential to the conservator. A re-installed *White Walls*, whether it delaminates or not, or to what extent, could be appreciated as 'an unsettling aftermath,' but one that critically acknowledges the 'already unsettled home' of its first incarnation.

Notes

1. The installation of *White Walls* at Galerie Lelong, New York began on 3 May 2007 and concluded at 9am on 8 May. The exhibition opened to the public at 10 am, one hour later. It closed on 16 June.

2. In planning for the installation, Goldsworthy had initially been concerned that the clay, when it fell, would pull paint off the walls. To help mitigate this, the walls were painted with water-resistant white emulsion paint in advance of installation. The first large tranche of wet clay fell overnight between 8 and 9 May. The short wall from which it fell was newly built. The other walls were either structural or else longer-standing partitions. However, the falling clay indiscriminately pulled paint and plaster from those too.
3. This is fairly unusual for Goldsworthy, there being few of his installations of any media that he currently feels can be re-installed without his participation or presence. The specific material used for *White Walls* was refined porcelain clay, originally from Cornwall, England, and acquired through a distributor in Massachusetts. Frequently for his clay installations, Goldsworthy uses raw, unprocessed clay, sourced locally. Using a refined commercial product for *White Walls* means that it could be more easily replaced – which Goldsworthy assumed would have to occur eventually.
4. Jacques Derrida elaborated his discussion on iterability in "Signature Event Context", a controversial paper that he first delivered in 1971 at a conference entitled *Communications*, and which he published in French in *Marges de la Philosophie* in 1972 (Paris: Minuit). The paper, which presented a critique of philosopher J.L. Austin's text "How to Do Things with Words" (1962), was also published in 1977 in *Glyph* 1 (translated by Samuel Weber and Jeffrey Mehlman). It has subsequently been published in *Limited Inc,* ed., G. Graff (Evanston: Northwestern University Press, 1988), alongside a summary of John R. Searle's 1977 defence of Austin, and Derrida's subsequent reply to that, "Limited Inc a b c…", (itself first published in *Glyph* 2, 1977).
5. Jacques Derrida, "Signature Event Context" in *Limited Inc*, ed., G. Graff (Evanston: Northwestern University Press, 1988) 61–62.
6. Martha Buskirk, "Locating the Intermittent Work of Art", presented at the Bonnefanten Museum, Maastricht, 11 May 2006, as part of the international, multi-institutional documentation project, *Inside Installations* (Video downloadable from *http://www.inside-installations.org/project/detail.php?r_id=240&ct= maastricht*). See also Jill Sterrett, presentation at Tate Modern, 22 March 2007, as part of "Shifting Practice, Shifting Roles? Artists Installations and The Museum" (*http://www.tate.org.uk/onlineevents/webcasts/shifting_practice_shifting_roles/default.jsp*).
7. Deconstruction is a term that Jacques Derrida coined in 1967, and is most closely (but not exclusively) associated with his literary criticism and philosophical writings. Difficult to define, it does not constitute a methodology or school of philosophy. David Allison described it as 'a project of critical thought' charged with the task of locating and 'tak[ing] apart' concepts and binaries that 'serve as the axioms or rules for . . . the unfolding of an entire epoch of metaphysics'. Jacques Derrida, *Speech and Phenomena and Other Essays on Husserl's Theory of Signs,* trans. David B. Allison (Evanston: Northwestern University Press, 1973).
8. Jon Ippolito first employed the term 'paradigm shift' with regard to the conservation of contemporary art in his Introduction to *Preserving the Immaterial,* a conference at the Solomon R. Guggenheim Museum, New York, March 2001, and the first public event organized by the Variable Media Network (or Variable Media *Initiative* as it

was then called). In his Introduction, Ippolito noted, 'the opportunity [. . .] is to craft a new collecting paradigm that is as radical as the art it hopes to preserve. The choice is ours: do we jettison our paradigm? Or our art?' The term 'paradigm shift' was first used by Thomas Kuhn in his book, *The Structure of Scientific Revolutions*, published in 1962. In his address at the above conference Ippolito appears to invoke Kuhn's concept, wherein a presiding system of assumptions, beliefs, values and practices gives way to change (transcripts of those proceedings are available for download at *http://www.variablemedia.net/e/preserving/html/var_pre_index.html*).

9. Alterity is the state or quality of being 'other'. As Galen A. Johnson explains, 'the term derives from the Latin *alteritas*, meaning "the state of being other or different; diversity, otherness". It does have its English derivations: alternate, alternative, alternation and alter ego. The term *alterité* is more common in French, and has as its antomyn *identité*'. See "Introduction: Alterity as a Reversibility", *Ontology and Alterity in Merleau-Ponty*, eds., Galen A. Johnson and Michael B. Smith (Evanston, Ill.: Northwestern University Press, 1990) xviii.

10. 'Variability' is typically associated with artworks comprising ephemeral materials or those that have no permanent physical constitution. It is taken to connote a range of aspects attributable to installation works of art: permissible material replacement outside of the original medium or substitution of items; changes in configuration from version to version and so on. It first appeared as a suffix to 'medium' – as in 'medium-variable' along with other suffixes '-specific' or '-independent', but more recently appears to have evolved from being a qualifier to a distinct condition that a work can exhibit – 'variability'. See Alain Depocas *et al., Permanence Through Change: The Variable Media Approach* (New York: Solomon R. Guggenheim Foundation, 2003) and Sterrett, 2007.

11. Nelson Goodman, *Languages of Art: An Approach to a Theory of Symbols* (Indianapolis: Bobbs-Merrill, 1968). Recently, the 'allographic' has been considered in relation to works of conceptual art: see Art & Language "Voices Off: Reflections on Contemporary Art", *Critical Inquiry*, 33, 2006, 125–126, and Kirk Pillow, "Did Goodman's Distinction Survive LeWitt?" *The Journal of Aesthetics and Art Criticism* 61:4, Fall 2003, 365–380. Regarding its application to installation art works, see Pip Laurenson, 'Authenticity, Change and Loss in the Conservation of Time-Based Media Installations', *Tate Papers*, Autumn 2006 (available at *http://www.tate.org.uk/research/tateresearch/tatepapers/06autumn/laurenson.htm*). (see also Stephen Davies, *Musical Works and Performances: A Philosophical Exploration*, 2004, published by Clarendon Press). For its application to digital artworks with a sound or music component, see John Roeder, 'Preserving Authentic Interactive Digital Artworks: Case Studies from the InterPARES Project' conference paper at ICHIM 04, Berlin (downloaded from *http://www.archimuse.com/publishing/ichim04/3185_Roeder.pdf*). For analogy to musical scoring vis-à-vis notation for new media artworks, see Richard Rinehart, "A System of Formal Notation for Scoring Works of Digital and Variable Media Art", 2003, p. 2(available for download from *http://bampfa.berkeley.edu/about_bampfa/formalnotation.pdf#search=%22richardrinehart%20pdf%22*).

12. Jacques Derrida, "Différance", *Margins of Philosophy*, trans. Alan Bass (Chicago: University of Chicago Press, 1982) 1–28.

13. Jacques Derrida, *Disseminations,* trans. Barbara Johnson (Chicago: University of Chicago Press, 1981) 168.
14. Thomas Baldwin, "Presence, Truth and Authenticity", *Derrida's Legacies: Literature and Philosophy*, eds., Robert Eaglestone and Simon Glendinning (London: Taylor & Francis, 2008) 110.
15. Jennifer Mundy, "Why/Why Not Replicate?" *Tate Papers: Special Issue on Replication*, 8, Autumn 2007 (available at *http://www.tate.org.uk/research/tateresearch/tatepapers/07autumn/mundy.htm*).
16. Tina Fiske, *Taking Stock: A Study of the Acquisition and Long Term Care of 'Non-Traditional' Contemporary Artworks by British Regional Collections 1979 – Present*, PhD thesis, unpublished, University of Glasgow 2004: Chapter Two. For an overview of the recent trajectories for literary translation studies, see Lawrence Venuti's "Translation Studies: an emerging discipline", *The Translation Studies Reader* (London: Routledge, 2004) 1–9.
17. Philip E. Lewis, 'The Measure of Translation Effects', *The Translation Studies Reader*, ed., Lawrence Venuti (London: Routledge, 2004) 256–275.
18. Lawrence Venuti, *The Translator's Invisibility: A History of Translation*, 2nd revised edn (London: Routledge, 2006) 18.
19. Derrida uses the term 'anethical' in the latter sections of *Limited Inc* (1988). He suggests that any given ethics will 'exclude, ignore, relegate to the margins other conditions no less essential to ethics in general . . .' (122). More recently, the concept of 'anethical' has been used by Shane Weller in relation to the literature of Samuel Beckett and Beckett as a 'self-translator'. See Weller's *Beckett, Literature and the Ethics of Alterity* (Houndsmill, Basingstoke: Palgrave Macmillian, 2006) 192–195.
20. Lewis proposed that his 'abusive fidelity' would serve a 'dual function – on the one hand, that of forcing the linguistic and conceptual system of which it is a dependent and on the other hand, of directing a critical thrust back towards the text that it translates and in relation to which it becomes a kind of unsettling aftermath (it is as if the translation sought to occupy the original's already unsettled home, and thereby, far from 'domesticating' it, to turn it into a place still more foreign to itself)' (Venuti, 2004:182).

Index

A
A:shiwi A:wan Museum and Heritage Center (AAMHC) (Mexico), 202, 204, 206
Aboriginal *see* indigenous
access, 3, 30, 67, 68, 70, 136, 141, 153, 154, 158, 197–209
accretion, 15, 132–3
    analysis/assessment, 132–3
    see also information value
ACM *see* Asian Civilisation Museum
'action'-based behaviour, 11
additions, 61, 88, 96, 158, 159, 160
advocacy planning theories, 180
aesthetic value, 25, 118, 174
    and conservation-restoration, 117
    and cultural heritage preservation, 116–18
    and 'finds,' 227
    and installation art, 227
Age of Enlightenment, xvi, 26, 30
Age of Reason, 74
age value, 32, 85, 117, 118, 119
    see also patina
'agency,' 214, 220
    human, 68, 183
AICCM *see* Australian Institute for the Conservation of Cultural Material
Aleppo Room, 152–4, 158, 159, 160
'allographic,' 65, 234–5, 240
alteration, 34, 50, 51, 52, 53–5, 57, 66, 71, 90, 126, 159–160
analytical methods, 117, 119
anthropology, 27, 113, 158
'anti-scrape,' 1–3
antiques, 126, 171
    renovation/restoration of, 42, 100
appearance, 21, 26, 28, 33, 51, 56, 84, 85, 88, 94, 97, 117, 126, 133, 134, 154, 157, 226–7

    disappearance and re-appearance, 235
archaeology, 15, 129–38, 169, 191, 227
    conservation aims, 27
    cultural significance, 145
    minimal intervention, 50
    'Rubbish Theory,' 169–70
    ruin reconstruction, 34
architects, 17, 27, 37, 42
architecture, 2, 78, 126, 222
    conservation aims, 27
    conservation of, 15, 63
    minimal intervention, 48
    see also restoration
Argan, Giulio C., 75, 114
art, 1, 15, 117, 118–19
    conservation aims, 27, 29
    contemporary museums, 223–8
    cultural constructs, 169–73
    cultural heritage preservation, 116–19
    installation, 210–22, 229–40
    international contexts, 76, 79
    minimal intervention, 47
    practical ethics, 64, 65
artefacts/artifacts, 27, 129, 130, 131–2, 134, 146, 188, 192, 206, 210, 212, 218, 219
    and authenticity, 60–5
installation art, 212–14, 216
    investigations, 132–3
    status of, 212–13
Arte Povera, 226
artist, 7, 64, 80, 100–12, 212, 213, 214, 215, 216, 217–18, 219, 227
    artist-restorer, 86, 87, 90, 101
    authority, 234
    instructions, 235–6
    intent, 85, 118, 230
    and value, 166
Ashley-Smith, Jonathan, xv, xvii, 174
Asian Civilisation Museum (ACM), 154
Australia, 34–5, 184–96
    Australian citizenship, 186

Australian Institute for the Conservation of Cultural Material (AICCM), 189
authenticity, 4, 20–2, 43, 61–3, 70, 71, 72, 78–81, 84–99, 105, 141, 145, 154, 163, 164, 167, 179, 182, 235
    and archaeology, 34, 37
    and the Berlin Aleppo Room, 158
    certificates of authenticity, 216
    and cleaning, 27
    and conservation, 167
    co-ordinates of, 64–5
    and culture, 153–4, 158, 170–1
    cultural interpretations, 85
    cultural significance of, 145
    definition of, 79, 80, 105
    and ethics, 61
    as an ideal, 163, 164
    and integrity, 73, 76, 77
    installation art, 216
    international contexts, 78, 79–81
    James Clifford, 170
    legal verification, 79
    material authenticity, 90, 114, 120
    Nara Conference on Authenticity, 79, 85
    nineteenth-century notions of, 86–87
    object-authenticity, 218
    practical ethics, 60–3, 64–5
    and principles, 73
    and restoration, 48, 86–9, 96–7
    and re-restorations, 96–7
    twentieth-century notions of, 87–90
    Venice Charter, 76
    and world heritage, 77, 79
    see also truth values
'authority,' 194, 199, 200, 207, 216, 234
    ancestral authority, 156
    and installations, 234
    and museums, 228
    Treaty of Waitangi, 186

Index

authorship, 62–64
auto-icons, 1–5
autographic art, 80, 234–5
Avrami, Erica, xvii

B

Baldwin, Thomas, 235
Barney, Matthew, 224–5
basket-weave pattern, 88–9
Bauerová, Zuzana, xvii
behaviour, 11–12, 14
  and chaos, 11
Bentham, Jeremy, 3, 10, 60
Berlin Aleppo Room, 152–4, 158, 159, 160
Berthollet, Claude-Louis, 104–5
biculturalism, 185–6, 187, 188, 191–2
binding media, 88
Birnbaum, Vojtech, 119
Blackmore, Susan, 9, 10
Boas, Franz, 139
Brajer, Isabelle, xvii
Brandi, Cesare, 7, 17–18, 32, 75–6, 77, 80, 95, 114, 121
Brera Pinacoteca, 106–7
British settlements in Australia, 185
building protection, 1–4
built heritage, 42, 77, 79, 181
Buren, Daniel, 212
Burra Charter, 7, 34–5, 78, 178–9
Buskirk, Martha, 226, 227, 232
Butts, David, 188

C

CAMA *see* Council of Australian Museum Associations
Cane, Simon, xvii
Canova, Antonio, xvi
capitoli, 47
Caple, Chris, xvii, 25–31, 49, 56
'Carlsberg Preparation,' 88
Carta italiana del restauro, 48
Cavalcaselle, Giovanni, 105, 108–9
caveats, 56, 68, 234
charters, 7, 16, 17, 20, 32–5, 37, 41, 42, 78
  creation, 7
  international contexts, 73, 75, 76–7, 78–9, 81–2
  planning theory, 178–9
  and ruin reconstruction, 33–5
China, 78, 82
Clavir, Miriam, xiv, xvi, xvii, 21, 164, 174

cleaning, 18, 26, 27, 28, 32, 47, 49, 50, 52, 57, 88, 96, 105, 107, 132–4, 141, 145, 152
Clifford, Helen, xvii
Clifford, James, 170–1
Codes of Ethics, 139–147
  see also guidelines
collaboration, 77, 78, 178, 191–2, 197, 198, 201, 219, 220
collections, 3, 14, 15, 20, 30, 130, 131, 141, 147, 175, 187, 191, 192, 198, 200–2, 205, 207
  archaeology, 135
  Code of Ethics for Museums, 66
  conservation aims, 30
  and continued collecting, 30
  deterioration, 141
  disposal, 173
  human interaction, 8
  management, 6, 16, 20
  principles, 77, 159
  and social/cultural constructs, 146, 166–7, 173
  value, 166, 167
  virtual, 30
Colonial Williamsburg, 35–7, 40, 41, 42
commoditized objects, 166–7, 171–3
communication, 2, 49, 136, 234
community involvement, 84, 207
  (*see also* public involvement, stakeholder participation)
complete compositions, 94
compromise, 56, 84, 96, 97, 141, 155, 157, 215
'conceptual value change,' 164
condition, conditions, 4, 18, 48, 51, 54, 61, 63, 95, 103, 107, 108, 129, 130, 153, 168, 203, 204, 211
  authenticity, 85, 86
  deterioration, 144
  environmental, 49, 108, 130, 134, 144, 153, 203, 231
  of integrity, 81
  Madonna di Foligno, 104
  political, 179
  and reconstruction, 42, 78
  replication, 132
  of silver, 127
  social, 180
  storage, 30, 134
condition documentation, reports, surveys, 19, 103, 106, 165
conferences, 12, 17, 18, 74–5, 77

conservation:
  aims, 25–30, 56
  archaeological, 129–37
  of carvings, 143, 192
  cultural significance, 139–49
  as cultural construct, 163–76
  Czechoslovak, 113–24
  definition, 141, 147, 174
  dilemmas, 1, 54
  disciplines, *see also* specialist conservation 14, 16, 17, 18, 27, 174, 190
  ethics, 61–70, 184–94
  globalisation of, 177–82
  history, 7
  installation art, 210–22, 223–8, 229–40
  as a local act, 179, 182–3
  and local knowledge, 178–9, 182
  and local policy, 178
  methodologies, 207, 211
  and minimal intervention, 47–57
  prehistory of, 8–10
  principles, 56, 73–83, 100–12, 150–62
  process of, 181
  profession, 7, 15, 20, 163, 164, 187, 189
  professional bodies, 192
  of ruins, 32–42
  as social practice, 135–6
  as social process, 61
  theory, 7, 14, 16, 32, 82, 163, 193
  values, 177–83
  see also restoration
conservator:
  aims and goals, 27, 54, 96, 141, 163–176, 192–3
  as arbiter of aesthetic value, 174
  and authenticity, 84, 97–8
  behaviour, 6–14, 201
  and 'convergence,' 19–21
  definition, 28, 190
  effect of actions, 55
  intervention, 52, 53, 156, 218
  perspectives on Codes of Ethics, 189–92
  principles and, 56, 116
  professional standing, 19–20
  role, 179, 218–19
  subjectivity of, 65
  in the Western world, 32, 73–4
consolidation processes, 153, 203
Consoni, Nicola, 108–9
consultation processes, 192, 202–3
  see also stakeholders' dialogue
consumption of objects, 125–6, 150, 151

242

Index

contemporary art, 115, 192, 219, 223–9, 233
Continuing Professional Development (CPD), 20
contractual agreements, 9
conventions, 33–5, 77, 177
  see also World Heritage Convention
Cooke the Younger, Henry, 102
copying processes, 80
  see also imitation behaviour; replicates
corrosion, 133, 134
Council of Australian Museum Associations (CAMA), 188–9
CPD see Continuing Professional Development
Cruikshank, Julie, 143
culture analysis, 150, 158–60
cultural constructs, 163–76, 201
commoditisation:
  of objects, 26, 166–7, 170, 171, 172, 173
  of culture, 166–7
cultural difference, 180, 193, 237
cultural dynamics, 150–62
cultural heritage, 61, 74, 79, 82, 116–19, 143, 164–5, 174–5, 187–9, 191
  Czechoslovak, 114–15, 120
  definition, 77, 78
  immovable, 177
  indigenous, 187, 188, 191
  integrity of, 81
  intangible, 143, 207
cultural environment, 174
cultural significance, 135, 136, 139–49, 187
  see also significance
culture:
  ethics, 185–7
  evolution, 8–9, 76
  living objects, 197–209
  living cultures, 197 – 209
  see also heritage conservation
Czechoslovak conservation-restoration, 113–24

D

damage, 1–2, 14, 15, 26, 34, 48, 52, 76, 89, 94, 95, 105, 130, 133, 134–5, 159, 204
da Vinci, Leonardo, 107
De Saussure, Ferdinand, 114–15
decision-making processes, 15, 56, 57, 63, 66, 69–70, 174, 179, 181, 182, 190, 191, 193, 227
minimal intervention, 55–6, 57
practical ethics, 63, 66
  see also process/processes
Deconstruction, 235, 236
decorative arts, 15, 18
degradation see deterioration
Demos think tank, 164–5
Denmark, 84–99
Derrida, Jacques, 232–8
  and the 'anethical,' 238, 241
desalination, 203
'desire lines' around sites, 39
destroying evidence, 15, 38–9, 132, 133
  cultural significance, 145
  minimal intervention, 52, 55
  reshaping, 134–5
deterioration, 28, 56, 65, 97, 146, 163
  archaeology, 130, 131, 132, 133, 134
  cultural significance, 144
  definition, 144
  desirable deterioration, 144
  installation art, 214
Devetsil art group, 115
Dia Art Foundation, 216
dialectic relations, 158, 160
digital economy, 68
digital media, 67–8, 215, 217
dilemmas, 1, 3, 70, 96, 212, 214, 236–7
  see also conservation dilemmas
disinfection, 133–4
display, 27, 28, 30, 90, 92, 97, 130, 131, 134, 152–9, 170, 173, 175, 210, 213, 218, 227
display equipment, 213–17
redisplay, 64, 65
documentation, 16, 37, 61, 62, 66–7, 68, 69, 101, 108, 113, 134, 153, 216–19, 235
  and early restoration of Raphael's paintings, 101, 108
  in the digital era, 190
  see also charters
Doerner, Max, 120
Drysdale, Laura, 152
durable objects, 169–70, 171
Dvorak, Max, 118
dyads, 114

E

Eastop, Dinah, xvi, xvii, 164, 169, 174
Edwards, Pietro, 47

eel net trap, 156–7, 159, 160
El Greco, 51
electrolytic reduction, 134
emulation, 215
  see also memes
Engum Church, 86–7
Enote, Jim, 204–208
epistemology, 61, 178
equipment, 80, 102, 120, 212, 214–15, 217–18
  dedicated/non-dedicated, 214
erroneous information, 38, 42, 61, 213
ethics, 6–24, 60–72, 139–148, 163, 174, 177, 179, 182, 184–96, 201
  of alterity, 233
  culture, 139, 140, 141
  fashion/tradition, 13–18
  minimal intervention, 54
  reconstruction, 38
  as sources of principles, 12–13
  translation ethics, 238
ethnographic conservation, 27, 63, 145, 207
'evaluation before removal,' 62–3
Evans, Arthur, 37, 38, 42
evidence, 15, 26–7, 34, 35, 38–9, 48, 50, 52, 53, 62, 77, 78, 81, 101, 133, 135, 145, 160
  archaeological, 36, 40–2, 103, 131, 132
  and authenticity, 145
  conservation aims, 26–7
  cultural significance, 139–40, 145
  documentary, 33
  empirical, 66
  excavations, 130–1, 132
  material/physical/tangible, 32, 73, 136, 139, 200, 201
  minimal intervention, 52, 55
  preservation of, 42
  reconstruction of, 35–7, 38–9, 41, 42
  reshaping, 134–5
  scientific, 56, 147
  technological, 135
evocative value, 37
evolution, 7, 8–10
excavations, 36, 37, 41, 129, 130–2, 136, 145, 202, 204, 205
  see also archaeology
examination, 15, 28, 133

F

'faithful-to-facts' restoration, 48
fallibility, 225

Index

fashion, 9–10, 12, 13–18, 125–6
   in conservation, 21
financial values, 166–8, 171–3
'finds,' 227–8
First Czechoslovak Republic (1918–1938), 114, 120
Fiske, Tina, xvii, 219
Fjenneslev Church, 97
Flavin, Dan, 211
Fox, Howard, 225–6
fragments, 50, 52, 101, 132
   of panel paintings, 101
   of wall paintings, 86, 90–3, 94, 96–7
   restoration of, 48
Freedom of Information Act (UK), 68
fumigation, 26
function, 64, 85, 118, 142, 158, 159, 190
   buildings, 36
   of conservation, 164, 165, 168, 169, 173
   of contemporary curator, 226
   functional artefacts/objects, 25, 27
   of museums, 30, 166
   Structuralism, 115–6
   Synthetic conservation, 119

G

gamelan, 154–6, 158–60
gene-culture co-evolution, 8
   and the concept of contractual agreement, 9
gene–meme co-evolution, 9–10, 12
   and prepared learning, 10
George Gustav Heye Center, 199
Giovannoni, Gustavo, 75
Global-local dialectic, 177
global-universals *see* universal application
Goldsworthy, Andy, 229–32, 235, 238
   at Galerie Lelong, 229–31
   White Walls, 230–41
Gonzalez-Torres, Felix, 213, 216–17, 220, 226
Goodman, Nelson, 233–4
guidelines:
   in China, 78
   and the condition of integrity, 81
   ethical, 11, 12, 33, 34, 35, 41, 63, 82
   gamelan, 154, 155
   heritage policy, 178

installation art, 216, 219
   restoration, 102, 108, 118
   universal, 14
   see also Codes of Ethics
Guyton de Morveau, 104–5

H

Habermas, Jürgen, 178
Hacquin, François-Toussaint, 104–5
   transfer technique, 105
Handler, Richard, 139–40
'hands-on' access, 141
'hands-on' skills, 218
Harding, Sarah, 140–1
Hassard, Frank, 20
Hawikku pottery, 202–8
Hedensted Church, 91–2, 94
Heidegger, Martin, 80
heritage conservation, 8, 18, 20, 73, 82, 114, 115, 116, 120, 164–5, 174, 177–83, 187–9, 191
   cultural significance, 139, 141, 143, 144–5
   in Czechoslavakia, 113–24
   international contexts, 73, 74–5, 76, 77–9, 81–2
   minimal intervention, 52
   socio-cultural contexts, 187–9
   structuralism, 116–19
   see also World Heritage Convention
heritage:
   archaeology, 130, 137
   built, 42, 74, 77
   collections, 147, 202, 204, 206, 220
   cultural significance, 139, 141, 143, 184–96
   definitions, 73, 77, 78
   future of, 175
   Indigenous, 187, 188, 193
   intangible, 79, 82, 130, 143, 207
   integrity, 81
   managers, 41
   objects, 52
   policies, 77, 187, 188
   restoration theory, 76
   tangible, 21, 82, 90
   values, 177–83
   World Heritage List, 85
historic accounts, 15–16
   Czechoslovakia, 114–15
   documentation, 27
   ethics, 7–10, 12–13, 185–7
historiographic accounts, 115–16

Hlobil, Ivo, 118, 119
Hoeniger, Cathleen, xvii
Holden, John, 164–5, 174
Holyoake, Manfred, 18–19
horizontal displacements, 39, 42

I

ICCROM *see* International Centre for the Study of the Restoration and Preservation of Cultural Property
ICOMOS *see* International Council on Monuments and Sites
identity, 141, 236, 237
   and authenticity, 62–3, 86
   and installations, 218, 233, 235
IIC *see* International Institute for Conservation of Historic and Artistic Works
Imitation, 80, 93, 117
imitation behaviour, 9–10, 80
   see also memes
'importance' designation, 140–1
   see also cultural significance
in-filling, 156, 157, 159
'in-painting,' 48, 108
   see also 're-painting'
inaccurate information *see* erroneous information
inauthenticity *see* authenticity
INCAA *see* International Network for the Conservation of Contemporary Art
inclusive conservation, 129, 136, 178, 183, 193
income, 19, 36
Indian National Trust for Art and Cultural Heritage (INTACH), 79
indigenous populations, 137, 146, 147, 178, 184–7, 188, 189, 191–2, 193, 197, 199, 201, 202, 207
   see also Native Americans
information:
   collection/transmission, 15, 26, 27, 38, 42, 67–9, 70, 130–1, 133, 134, 174, 215
   through cleaning, 28
   loss of, 29, 52
   museum as source of, 225
   preservation, 30
   see also erroneous information
Ingold, Timothy, 158
inheritance, 3, 9, 61
   see also evolution

Index

installations, 210–22, 224–5, 230–41
   of candy, 213, 216–17
   and collaborators, 217–18
   and dedicated display equipment, 214
INTACH *see* Indian National Trust for Art and Cultural Heritage
intangibility, 9, 20–1, 73, 78, 79, 80, 82, 85, 130, 135, 181, 190, 200–1, 207
   archaeology, 135, 136
   cultural significance, 135, 136, 139, 143, 168
   intangible values, 84, 97, 186, 191
   intangible heritage, 207
   international contexts, 73, 77–80, 82
integrity, 11, 73, 76, 77, 81–2, 131, 141, 145, 155–6, 157, 159, 190, 207, 226
   aesthetic, 81
   archaeology, 131, 134, 135
   conceptual, 155, 159
   cultural dynamics, 155–6, 157, 159–60
   cultural significance, 145
   excavated objects, 131
   historical, 81
   intellectual views, 12
   material, 135, 156, 160
   and monuments, 82, 117
   physical, 61, 131, 135, 158, 160
   and reconstruction, 42
intention, 62–4, 216–17, 234–5
   and authenticity, 85
   of conservation, 26, 54, 131
   of conservator/restorer, 47
inter-institutional communication, 67, 68
International Centre for the Study of the Restoration and Preservation of Cultural Property (ICCROM), 76–7, 80, 174
international contexts, 73–83
international conventions, 33–5
international legislation and guidelines, 33, 177
International Council on Monuments and Sites (ICOMOS), 34, 38–9, 77, 78, 177–8
International Institute for Conservation of Historic and Artistic Works (IIC), 13, 17, 146, 152, 189

The Object in Context: Crossing Conservation Boundaries Munich Conference 2006, 146, 151
International Network for the Conservation of Contemporary Art (INCCA), 219
interpretation, 64, 87, 95, 113, 116, 117, 129, 132, 190, 199, 202–3, 216–17
   of authenticity, 20–1, 85–6, 97
   of Codes of Ethics, 193
   cultural significance, 139–40
   curatorial, 216
   reconstruction, 34, 35
   restoration, 134–5
   of sites, 40, 81
   subjectivity, 139
   of values, 193
intervention, 47–59, 62, 63, 66, 82, 90, 95, 96, 97, 102, 113, 117–19, 134, 136, 151, 152, 154, 158, 164, 182, 218
   archaeology, 133–4
   by artists, 210, 230
   The Berlin Aleppo Room, 153
   cultural dynamics, 153–4, 155, 156, 157
   practical ethics, 62–3
   as social process, 150
   visibility of, 42, 105
intra-institutional communication, 67, 68
intrusions, 52–3
'investigation before intervention,' 62–3
investigations, 28, 29, 30, 131–3, 136, 151
   artefacts, 132–3
   practical ethics, 62–3
   scientific, 120
'involvement'-based behaviour, 11
Ise Shrine, 80–1
iteration, interability, 65, 219, 229–40

J
Jakobson, Roman, 115–16
Japan, 33, 36, 39, 78–9, 80–1, 82
Jedrzejewska, Hanna, 17–18
Jokilehto, Jukka, xvii, 85
Jones-Amin, Holly, 154–5, 161

K
Kaminitz, Marian A., xvii, 146
Kemp, Jonathan, xvii

Kennedy, Tom, 202–3
Knossos, 37, 38, 40, 42
knowledge management, 67–8, 69
   see also information collection/transmission
Kornerup, Jacob, 87
Kosuth, Joseph, 216
Kramar, Vincenc, 119–20
Kunstwollen, 79–80

L
lacquers, 14
lacunae, 88, 94, 95
Ladd, Edmund J., 144
landscape values, disruption of, 39
language, 4, 9, 49, 113–16, 121, 150, 152, 159, 237
   barriers of, 18
   of codes of practice, 193
   of heritage conservation, 177
   as intangible heritage, 130, 197, 198
   'The Language of Conservation,' 164
   of professional practice, 179
Laurenson, Pip, 65, 234
Le Brun, Jean-Baptiste-Pierre, 103
Leigh, David, 141, 144
Lewis, Philip, 236, 237
lime casein colours, 88
limitations, 8
   of the conservation lexicon, 164
   and minimal intervention, 47–9, 56
   of reconstruction, 41
limits:
   and conservation/restoration, 16, 18, 47–9, 120
   and installation art, 211
Lind, Egmont, 87, 88–9
linguistics, 113–16
literary theory, 115–16
living culture/objects, 135–6, 197–209
Lloyd-George, David, 19
loss:, 11, 12, 52, 55, 61, 66, 97, 103, 105, 108, 132, 156, 214, 215
   'balanced meaning loss,' 57
   installations, 214
   'minimal meaning loss,' 56
   minimization, 61
   value, 172
Louvre, 102–3, 104, 106
low status artifacts, 212–13

Index

M
Mabo judgement, 186–7
MAC *see* Maori Advisory Committee
McCracken, Grant, 125–6, 127
machinami charter, 78–9
Madonna del Cardellino, 101
Madonna di Foligno, 103–5
'maintaining identity,' 62–3
maintenance, 9, 21, 25–6, 28, 75, 155, 160
   costs, 41
   of communities, cultures, 198, 200, 201
   ethics, 9, 62–3
   manufacture, 14–15, 20, 27, 131, 133
   of silver, 125
Māori, 26, 156–7, 160, 184–196
   Advisory Committee (MAC), 157
marble, 55
Marriage of the Virgin, 105–8
Material, materials, 4, 9, 14, 15, 20, 21, 26, 42, 47, 50, 53, 54, 61, 63, 64, 66, 75, 79, 80, 81, 82, 85, 90–8, 102, 104, 113, 125, 129–138, 143, 157, 166, 167, 184, 204, 210, 214, 216, 224
   change, 136
   cultural, 184–5, 202, 203, 205, 208
   damage, deterioration, 14, 133, 144
   ephemeral, transient, 226, 227, 228
   and ethics/principles, 16, 62, 77, 129, 139, 141, 190
   historic, 73, 78
   inorganic, 130
   interpetation of, 129
   organic, 27, 55, 130
   original, 60, 63, 73, 100, 105, 109
   preservation, 84, 134
   preventive conservation, 132, 153, 193
   properties, 14–15, 133
   and reconstruction, 34–5
   and restoration, 20, 26, 34–5, 134
   storage, 134
   substitution, 158
   value, 101, 136, 160, 169
material additions: object stability, 133

material authenticity, 63, 64, 65, 70, 84, 89, 90–3, 95–8, 114, 120
material culture, 126, 143, 146, 150, 160, 168, 175, 185, 189, 199, 200
material culture analysis, 150–62
material evidence, 32, 73, 136
   see also objects
material integrity *see* integrity
material movement, 169–73
material remains *see* archaeology
materiality, 210, 219
Matero, Frank, 61, 189
meaning-loss, 55–7
mechanical properties, 54
Medieval wall paintings, 84–99
media art, 64, 65
memes, 9–10, 16, 17, 21
methodology, methodologies:
   of archaeology, 129
   of conservation, 62, 82, 88, 89, 94, 114, 117, 199, 207
   values, relation to, 179
Miller, Daniel, 158
Mingwei, Lee, 214, 215
Minimal/minimum intervention, 6, 9, 18, 19, 32, 47–59, 90, 106, 108, 109, 136, 153, 156, 157, 159
   approaches, 18
   and 'balanced meaning-loss,' 55–7
   cultural dynamics, 156
   definition, 49, 52–3
   treatment degrees, 49–51
minority groups, 140
mirrored panels, 212
Mithlo, Nancy Marie, 144
mixed status, 212–13
modern society, 8, 26–7, 97
modifications, 52, 53–4, 55, 131
Molotch, Harvey, xvii
Molteni, Giuseppe, 106–8
monuments, 2, 7, 26, 48, 77, 82, 117, 118–19, 201
   Congresses, 76–7
   minimal intervention, 48
   and principles, 76, 78
   reconstruction, 85
   restoration, 48, 74
   values, 118
Morris, William, 1–4, 73, 74, 75
moveable cultural property
   see collections, cultural heritage, museums
Mukarovsky, Jan, 115–16, 117, 118, 119

Muñoz-Viñas, Salvador, xvii, 7, 164, 174, 179
Murdock, George P., 9
museums, 3–4, 20, 21, 26–7, 63, 131, 135, 143, 144, 146, 163–75, 186, 187, 188, 191, 192, 197–209, 219, 220
   conservation aims, 26–7, 29, 30
   conservation practices, 146
   contemporary art, 223–8
   cultural and social constructs, 147, 166–73
   cultural dynamics, 154–6
   and ethics/principles, 66, 77
   function, 30
   and repatriation, 29
Museum für Islamische Kunst (Museum of Islamic Art), Berlin, 152

N
Nam June Paik, 216
Napoleon, 103–4
Nara, Japan, 7, 33, 36, 79
Nara Document on Authenticity (1994), 7, 79, 85
narrative, 143, 180, 182, 186, 226
   conservation as archaic, 174
National Museum of the American Indian (NMAI), 197–202, 204, 207
National Park Service (NPS), 41
national symbolic value, 35–6
Native Americans, 26, 197–202, 204
Native Title Amendment Act, 186–7
'natural world,' 26–7
negative intervention/modification, 52–3, 57
negative views of patina, 126–7
neglect, 1, 2, 9, 11, 81, 108
'neglect'-based behaviour, 11
net traps, 156–7, 159, 160
new media art, 64, 65
New Zealand, 156, 184–96
New Zealand Conservators of Cultural Material (ZNCCM), 184, 189
NMAI *see* National Museum of the American Indian
non-art–culture zones, 171, 173
non-governmental organizations (NGOs), 74–5
'non-minimal' intervention, 49–50

246

Index

non-objects related tasks, 19
noticeable restoration treatment, 53–5
NPS (National Park Service), 41

O

objectives, 63, 94–95, 134, 189, 198
objectivity, 49–51, 97, 116, 127, 165, 178, 225, 226, 228
objects, 3, 6–10, 13–15, 19, 25–31, 47, 51–5, 61, 63–5, 70, 76, 77, 84, 90, 101, 125, 126, 127, 129–38, 140, 142–7, 152, 156–8, 163–175, 200–1, 205, 206, 207
   and agency, 214
   meanings, 57, 85, 198
   sacred, 144, 187
   treatment, 19, 21
   value, values, 160, 184
   Victoria & Albert Museum, 14
   see also material evidence
'old' objects see age value; patina
orchestras, 154–6, 158–9
ordering, 158–60
   of information, 67
'original-state,' 3–4, 48, 51, 63–5, 86, 105–7, 117, 151, 154, 159, 160
   excavated objects, 131

P

paintings, 4, 47, 48, 50, 51, 100–12
   authenticity, 86, 88, 90–7
   cave paintings, 26
   cleaning, 32
   conservation, 27, 53, 100–12, 217
   wall paintings, 84–99
   see also re-painting
palimpsests, 90–1, 98
passive conservation, 156
   see also preventive conservation
patina, 32, 52, 125–8, 136, 144, 213
   and 'bright cut' engraving, 127
   consumption 'arenas,' 125–6
   definition of, 126
perception, 85, 97, 114, 117, 120, 154, 173, 174
performance, 36, 81, 154–5, 158, 159, 215, 219, 231, 233, 234
performance art, 210
performance indicators, 19
Philippot, Paul, 77, 80

Phnom Penh Prison, 165, 166
photography, 220, 224
physical integrity see integrity
physical change see material change
Picault, Robert, 103–4, 105
planning, 82, 153, 201
   conservation as form of, 182
planning theories, 178–9, 180
'play of tropes,' 154
polemics, 47, 118–19
political histories of Australia and New Zealand, 186–7
politics, 21, 119, 182, 210
   of conservation, 180
   employment, 6
postmodern era, 167–8, 177–9, 182
pottery, 50, 202–9
practical ethics, 60–72
Prague, 115–16, 117, 119
prehistory of conservation, 6–10, 12–13
'prepared learning' bias, 10
presence, 197, 228, 232, 234, 236–7
   artists' presence, 218
presentation, 84, 85, 154, 201
   aesthetic, 116
   of objects, 61
   of sites, 37
   virtual, 220
   of wall paintings, 97
preservation, 28, 29, 30, 61, 73, 78, 84, 90, 97, 101, 102, 105, 106, 113–19, 130, 141, 144, 145
   access and, 3, 141, 147, 190, 200, 207
   of buildings, 75
   and conservation, 136, 147, 151, 165
   community-based, 178
   cultural heritage, 116–19, 189
   cultural significance, 144–5
   Eigil Rothe's concept of, 88
   of evidence, 42
   installations, 215, 237
   site preservation, 37
   preservation status, 187
   as supreme principle, 84
   and Zuni culture, 144
preventive conservation, 17, 20, 28, 53, 84, 132, 134, 141, 153, 154, 157, 214, 227
   cultural dynamics, 157, 158–9
   cultural significance, 141
   of stored material, 134

principles, 3, 7, 30, 56, 121, 129, 146–7, 163, 179–80, 183, 190, 192
   and cultural significance, 139
   definition, 74, 129, 154
   of dialectics, 116
   restoration, 113
   and the rise of empiricism, 102–3
   and ruins, 32–46
   Structuralist, 116, 118
   and sustainability, 181
principles and conservation, 6, 8, 10, 49, 56, 61, 73–83, 84, 100–12, 114, 116, 119, 121, 129, 136, 141, 150–62, 163, 177, 178, 179–80, 182
process, processes:
   aesthetic, 88, 96, 98
   assessment, 78
   cleaning, 96
   conservation as social/cultural process, 61, 84, 129–38, 146, 150
   creative, 79–80, 85
   deterioration, 144
   integrating, 95
   material, 62
   of physical change, 164
   planning and management, 82
   of reconstruction, restoration, 36, 64, 81
   research, 36
   reversible, 13
   social, 63, 70
   stabilisation, 28
   see also decision-making processes
public archaeology, 130
public involvement, 18, 136
   archaeology, 130, 135–6
publications, 7–8, 16–18, 26, 84, 85, 199, 202

Q

qualitative/quantitative indicators, 165
qualitative judgement of authenticity, 80
Quam, Curtis, 204, 205

R

Raphael, 100–12
   Canigiani Holy Family, 101
   Cartoons, 101–2
   Madonna di Foligno, 103–5

Index

Raphael (*Continued*)
   Marriage of the Virgin, 106
   St. Michael Altarpiece, 103
rarity value, 8, 141–2, 166
Rauschenberg, Robert, 226
re-assembly of objects, 25
re-installation, 216, 218, 235, 236, 237
   see also installation
re-interpretations, 135
re-ordering, 158–9
re-painting, 26, 86, 88, 98, 104
re-restorations, 96–7
re-shaping, 134–5, 159
re-tuning, 155–6
'real evidence,' 145
Real, William, 215
reassembly, 25, 34, 102, 153
Reconciliation movement, 187–9
reconstruction, 66, 76, 81, 85, 86, 87, 88, 97, 98, 101, 117, 134, 153, 158
   arguments against, 37–41
   ceremonial, 80
   and continuing function, 36
   costs, 40–1
   definition/interpretation, 34–5
   and 'desire lines' around sites, 39
   and education, 36
   ethics, 38
   justification, 35–7
   post-war reconstruction, 76, 77, 119
   and practical ethics, 34, 37
   and preservation, 37, 42
   principles, 33–35
   're-use' reconstruction, 36
   ruins, 32–46
   and tourism, 36
records, 16, 30, 42, 61, 66–7, 68–9, 106, 204
   see also documentation
Rée, Jonathan, xvii
reintegration, 18
religion, 10, 12–13, 205
   see also sacred views
repatriation, 29
'replacement' pieces, 14, 81, 212, 217, 227
replicates, 80, 214, 215, 219
reported practice, 150–62
research, 28, 30, 36, 38, 50, 62, 66–7, 75, 164, 178, 179, 181, 199
   archaeological, 103
   and installation art, 216, 218

scientific, 104
research collections, 165
research value, 166
responsibility, 4, 69, 107, 118, 132, 174, 177, 186, 198, 226
restoration, 1–4, 7, 8, 9, 10, 11, 16, 18, 20, 28, 32, 35, 42, 63, 66, 73, 81, 82, 84–6, 119–20, 146, 152, 157
   Athens Conference 1931, 74–5
   and authenticity, 61, 96–7, 145
   Cesare Brandi, 75–7
   Colonial Williamsburg, 41
   and conservation principles, 73
   as conservation process, 64
   criteria, 42
   Czechoslovakia, 113–24
   early debates about, 74
   early 20th century, 87–9
   ethics, 14
   international contexts, 35, 73–7
   interpretation, 134–5
   'Japanese restoration,' 80
   like-with-like, 21
   minimal intervention, 47–8, 53–5
   Nobleman with his Hand on his Chest, El Greco, 51
   paintings by Raphael, 100–12
   vs. reconstruction, 34, 35
   re-restoration, 96–7
   ruins, 33
   techniques, 14
   William Morris and, 74
   see also conservation
'restrained conservation,' 48
retouching, 4, 48, 94–5, 109, 153
   methodology, 88–9
   'neutral retouching,' 152
   techniques, 48, 94–5, 96, 107
   values, 120
revelation, 28, 29, 30
reversibility, re-treatability, 6, 13–14, 18, 32, 49, 63, 133, 141, 157, 179, 182
revision control *see* version control
Richter, Vaclav, 119
RIP Triangles, 28–9, 30
risk, 4, 79, 107, 132, 133, 135, 156, 159, 165
risk assessments, 12, 20, 142
   artefact investigations, 133
   excavation, 132
risk aversion, 12
risk management, 12
rock art conservation, 191

Roeser, Mathias Bartholomäus, 105
Rothe, Eigil, 87–8
Rubbish Theory, 169–70
Ruhemann, Helmut, 48, 51
rupture, 233, 234
Ruskin, John, 1–3, 7, 48, 74, 75

S

Sacred, 74, 155, 187
   archaeology, 135
   cultural significance, 142, 144
   pictures, 51
   power, 155
   value of, 21
Salish communities, 144–5
salvage excavations, 37
'Scatter Art,' 211–12
scholarly knowledge management, 67–8
Schwed, J.M., 152–4
science, 55, 104, 191
   conservation and, 136–7, 146, 164, 168, 174, 178, 218, 227
   as methodology, 61, 70, 84, 90, 114, 116, 120, 129–38
   and minimal intervention, 50–1
   objectivity, 50–1, 127, 179
scientific conservation, 48
scientific definition of deterioration, 144
scientific developments, 30
   authenticity, 21, 61
scientific evidence, 56, 147
scientific obligations, 41, 42
scientific paradigm, 10, 21
scientific principles, 26
scientific values, 139
Scott, Marcelle, xvii
Secco-Suardo, Giuseppe, 107
Second International Congress of Architects and Technicians of Historic Monuments, 76
significance, 2, 4, 13, 75, 80, 81, 135, 179, 185
   archaeological, 131
   of built heritage, 181
   as decision-making tool, 193
   intangible, 135, 136
   see also cultural significance
signifiers, 114–15
silver, 13–14, 125–7
site interpretation, distortions of, 40

Index

Sixth International Congress of Architects, 48
Slansky, Bohuslav, 119–21
soaking, 133
social constructs, 141, 147
social difference, 178
society, 25–7, 28, 30, 76, 97, 126, 135–6, 141, 165, 167, 168, 175, 177, 178, 179, 181, 190, 191
   Australian citizenships, 186–7
   Czechoslovakian, 113, 114, 116
   and cultural significance, 139–140
   cultural traditions, 79
   museums and, 26–7
   New Zealand, 186
   values, 166
sooty particles, 203
Sourek, Karel, 118–19
souvenirs, 97
specialist conservation, 14–15, 20, 36, 55, 76–7, 102, 105, 130, 132, 134, 136, 191, 217, 219
   archaeology, 130, 132, 134, 136
   artefact investigations, 132–3
speed variability, 223–4, 228
spiritual practices, 79, 154–6
spiritual qualities, significance, values, 78, 79, 118, 135, 139, 154–6, 159, 186, 187, 189
spiritual views, 12–13, 21
   see also sacred views
'springs of action,' 10–12
stakeholders, 66, 82, 178–180, 182, 191
   dialogue/participation, 60
Staniforth, Sarah, 141
Stanley-Price, Nicholas, xvii, 25
status:
   artifact, 212–214, 216
   authenticity, 85
   of conservators, 19–20
   Indigenous people, 186
   literary translation, 237
   objects, 66
   patina, 126
   taonga, 187
Sterrett, Jill, xvii, 232
stone, 2, 53, 65, 135, 192
Stone Age behaviour, 8–9
storage, 26, 27, 30, 134, 156, 157, 159, 203, 210, 215
   of information, 69

Structuralism, 114–20
subjectivity, 50–1, 237
'success'-based behaviour, 11
support structures, 91–2, 104–7, 132, 151, 157
sustainability, 20, 177–83
   of cultures, 200–1
   of skills, 21
symbolic material culture, 143
symbolic meaning, 217
symbolic values, 35–6, 84
symbols, 26, 27
'Synthetic method of monument preservation,' 117, 118–19
synthetic polymers, 133
synthetic resin, 54
Sze, Sarah, 224

T

Tagish elders *see* Tlingit societies
'taken-for-granted' concepts, 151, 152
Tan, H., 154
tangibility *see* intangibility
taonga, 156, 157, 184, 186, 187–8, 192
Taunay, Nicolas-Antoine, 105
technical specialist, 21t
technology, 4, 14, 28, 43, 55, 67, 125, 134, 150, 215
Tee, A., 154
tension, 177, 179–80, 214, 219, 223
terra nullius, 185, 186
terracotta pots, 50
tethering, 233–4
   untethering, 61
think tank, 164–5
Thompson, Michael, 169–70, 171
threatened sites, 130
Throsby, David, 181
Thorvaldsen, Bertel, xvi
time, 6, 9, 10, 14, 19, 28, 32, 42, 50, 52, 54, 62, 76, 80, 85, 125, 126, 142, 179, 180, 215, 216, 223–5, 227, 228
'time-based media,' 65, 215, 234
time-based productions, 64
Tirsted Church, 96–7
Tlingit societies, 142–3
tools, 8–9, 26
Torres Strait Islander peoples, 185, 187, 188, 189
tourism promotion, 36
tradition, traditions, 12, 26, 73, 76, 79, 86, 126, 141, 142, 143, 205, 207, 225

conservation, 27–8, 64, 65, 84, 85, 86, 87, 90, 155, 182, 219, 226
cultural significance, 79, 80, 81, 143, 144, 145, 197
   in ethics, 13–18
   international contexts, 79
   museums, art, 166, 168, 214, 216, 224, 227
   skills, 20, 78
   traditional Zuni pottery, 202, 205, 208
transfer, 169, 170–3
   conservation procedure, 91, 92, 103, 104–5, 106, 107
   of knowledge, 67, 69
transient objects, 169–70
transient materials, 226, 227, 228
translation, 4, 12, 120
   literary, 233, 236–7
translation ethics, 237–8
tratteggio retouching, 88–9
Trinity of Saints, Raphael Sanzio, Chapel of San Severo, Perugia, 108–9
trust, 173, 174, 225–6
truth, 63, 73, 79, 80, 81, 86, 96–7, 98, 140, 145, 160, 163, 164, 167, 174, 175, 191
   and collections/museums, 226, 228
   in relation to 'principle,' 151, 154
   'truth values,' 60–1, 67
   see also authenticity
tuning, 155–6, 158, 160

U

UN *see* United Nations
'underpinning,' 156, 160, 218
United Nations Educational, Scientific and Cultural Organization (UNESCO), 17, 33, 51, 74–8, 177–8
United Nations (UN), 34, 74–5
unity, 15, 75, 80, 81, 85, 117–18, 119
universal, universals, 9, 16, 20–1, 177–8, 179, 190–1
   cultural heritage preservation, 147, 174, 180, 182, 183
   theory, 16
universal ethics, principles, 19, 76, 78, 82, 179, 189, 190
   charters, guidelines, 14, 35
unnoticeable object alterations, 54–5

Index

V
'validation'-based behaviour, 11, 12
value, values, 8, 13, 15, 25, 32, 35–6, 49, 51, 56, 60, 62, 63, 75, 76, 79, 80, 81, 82, 84, 85, 97, 100, 101, 109, 117, 118, 120, 125, 126, 135, 139, 140, 146, 147, 159, 160, 163–76, 177–83, 184, 186, 187, 188, 189, 190, 191, 192, 193, 198, 200, 201, 205, 207, 212, 226, 232, 236, 237
  cultural constructs, 164–5, 166–8, 169–73
  evocative, 37
  historic, 85, 91, 118, 119, 168
  intangible, 21, 84, 97, 191
  landscape, 39, 41
  museum, 144, 146
  original, 9
  patina, 125, 126, 136
  truth-value, 60, 65, 67
variability, 113, 216, 227, 228, 233–4
variable media, 63, 227
Variable Media Initiative, 216
variable speeds, 223–4, 227
varnish, 27, 50–1, 52, 54, 104, 107, 152, 158
Vasari, Giorgio, 101
'velvet' patina, 126–7
Venice, Italy, 33, 47
Venice Charter, 7, 17, 34–5, 76–8, 81–2, 177
Venuti, Lawrence, 236–7
version control, 68–9
vertical displacements, 39, 42
Vertue, George, 102
Victoria and Albert Museum, London, 14, 65
Villers, Caroline, 49, 56
Vincent, François-André, 105
Viollet-le-Duc, Eugène-Emanuel, 7, 47–9, 86–7
virtual, virtuality, 64, 220, 226
  access, 201
  collections, 30
  realities, 43
visceral views, 12
Vrigsted Church, 90–1, 93

W
Wagner, Vaclav, 116–18
Wain, Alison, 190
Waitangi treaty, 185–6, 188, 191
Walton, Parry, 102
'wantonness'-based behaviour, 11
wealth, 10, 12–13, 30, 100, 126
weaving conservation, 144–5
West, Richard W., 197–209, xvii
Wharton, Glenn, xvii
White Walls, 230–41
wholeness, 81, 158
Wik judgement, 186–7
Wikipedia, 69
wood panel conservation, 104, 106, 107, 152–4
workmanship, 79, 125
'works of art' *see* art
World Heritage Convention, 33–4, 41, 77, 79, 81, 177
World Wars, 74, 75–6, 77, 113, 152

Y
Yatsattie, Eileen, 202, 204, 205

Z
Zanchi, Antonio, 107
Zdenek, Wirth, 119
ZNCCM *see* New Zealand Conservators of Cultural Material
Zuni people, 144, 202–7